ROUTLEDGE LIBRARY EDITIONS:
URBAN PLANNING

Volume 23

URBAN LAND AND
PROPERTY MARKETS IN
THE UNITED KINGDOM

URBAN LAND AND PROPERTY MARKETS IN THE UNITED KINGDOM

RICHARD H. WILLIAMS AND BARRY WOOD

Routledge
Taylor & Francis Group

LONDON AND NEW YORK

First published in 1994 by UCL Press

This edition first published in 2018
by Routledge
2 Park Square, Milton Park, Abingdon, Oxon OX14 4RN

and by Routledge
711 Third Avenue, New York, NY 10017

Routledge is an imprint of the Taylor & Francis Group, an informa business

British Library Cataloguing in Publication Data
A catalogue record for this book is available from the British Library

ISBN: 978-1-138-49611-8 (Set)
ISBN: 978-1-351-02214-9 (Set) (ebk)
ISBN: 978-1-138-49035-2 (Volume 23) (hbk)
ISBN: 978-1-138-49039-0 (Volume 23) (pbk)
ISBN: 978-1-351-03542-2 (Volume 23) (ebk)

Publisher's Note
The publisher has gone to great lengths to ensure the quality of this reprint but
points out that some imperfections in the original copies may be apparent.

Disclaimer
The publisher has made every effort to trace copyright holders and would welcome
correspondence from those they have been unable to trace.

Urban land and property markets in the United Kingdom

Richard H. Williams
Barry Wood
University of Newcastle upon Tyne

UCL
PRESS

First published in 1994 by UCL Press

UCL Press Limited
University College London
Gower Street
London WC1E 6BT

The name of University College London (UCL) is a registered
trade mark used by UCL Press with the consent of the owner.

ISBN:
1-85728-048-2 HB

British Library Cataloguing-in-Publication Data
A catalogue record for this book
is available from the British Library.

Typeset in Times Roman.
Printed and bound by
Biddles Ltd, King's Lynn and Guildford, England.

CONTENTS

PREFACE

This account of the urban land and property markets of the United Kingdom is based on a report prepared in 1991 by the authors, together with Agnette Linn, as part of the five-country EuProMa project described in the Foreword. We would like to place on record our appreciation of the contribution made by Agnette Linn during her time as a research associate working on this project.

We are grateful to our colleagues in the Department of Town and Country Planning, University of Newcastle upon Tyne, for contributing their advice and assistance. Particular mention must be made of Rose Gilroy for her advice on aspects of housing policy and for setting up the Falstone Walk case study (Ch. 9.3), Colin Wymer for his continuing willingness to sort out all our computing problems and keep the word processors flowing, and Peter Sanderson for preparing the figures.

Assistance in preparation of the case studies was given by various firms of estate agents, chartered surveyors, developers and local authority planning departments. In particular, assistance from the following is gratefully acknowledged: Stockley Park Management Ltd (Ch. 6.1), Tyne and Wear Development Corporation and Dysart Developments (Ch. 6.2), Newcastle upon Tyne City Council (Chs 6.2 and 9.3), Aberdeen City Council, Grampian Regional Council and Headland Homes (Ch. 6.3), Hertsmere Borough Council (Ch. 9.2), and Wimpey Homes (Ch. 9.3). All kindly gave time to provide information not available from documentary sources.

Most of the contents of Part II ("The urban land market") apply also to the operations of the urban property market. No distinction is recognized in the Town and Country Planning Acts, for example, between regulating first urban use and any subsequent development. Consequently, a comprehensive account is given in Part II, and Part III concentrates only on those topics for which specific treatment of the property market is necessary.

An unusual feature of this volume is the inclusion as an Annex of the report in German entitled *Immobilienmärkte in Grossbritannien* (Real estate markets in Great Britain) compiled in 1991 by Ulrich Pfeiffer, Volker Gillhaus and Boris Braun of Empirica Forschung, Bonn, for the German Federal Ministry of Regional Planning, Building and Town Planning (Bundesbauministerium). We are grateful to the Ministry for the suggestion that the research report they commissioned should be given wider circulation by incorporation in this volume, and to the DIrector of Empirica, Ulrich Pfeiffer, for agreeing to make their research available for publication in this manner and for making the necessary editorial changes. We hope that, at least for readers of German, this will offer an alternative perspective, looking at Britain from the outside, as a contrast to that offered by ourselves. We remain wholly responsible, however, for the English text.

RICHARD WILLIAMS   BARRY WOOD

Newcastle upon Tyne, SEPTEMBER 1993

ABBREVIATIONS AND ACRONYMS

ARCUK Architects Registration Council of the United Kingdom
ARICS Associate of the Royal Institution of Chartered Surveyors
BR British Rail
BURA British Urban Regeneration Association
CBD central business district
CG City Grant
CGT Capital Gains Tax
CIA Commercial Improvement Area
CPO Compulsory Purchase Order
DC development control
DIY do-it-yourself
DLG Derelict Land Grant
DOE Department of the Environment
DOE(NI) Department of the Environment (Northern Ireland)
EA Environmental Assessment
EAGGF European Agricultural Guidance and Guarantee Fund
EC European Community
EIA Environmental Impact Assessment
EIP Examination in Public
ERDF European Regional Development Fund
ERM Exchange Rate Mechanism
ESF European Social Fund
EZ Enterprise Zone
FRICS Fellow of the Royal Institution of Chartered Surveyors
FRTPI Fellow of the Royal Town Planning Institute
GDO General Development Order
GIA General Improvement Area (housing)
GNP Gross national product
ha hectare
HAA Housing Action Area
HAT Housing Action Trust
HMIP Her Majesty's Inspectorate of Pollution
HRA Housing Renewal Area
IHT Inheritance Tax

IA Improvement Area (industry and commerce)
IIA Industrial Improvement Area
IOH Institute of Housing
ISVA Incorporated Society of Valuers and Auctioneers
km kilometre
LA local authority
LBCA Listed Buildings and Conservation Areas
LC licensed conveyancer
LPA local planning authority
m metre
M million
MIOH Member of the Institute of Housing
MP Member of Parliament
MPG Minerals Policy Guidance (DOE)
MRTPI Member of the Royal Town Planning Institute
OMA Outer Metropolitan Area
PLI Public Local Inquiry
PPG Planning Policy Guidance (DOE)
QUANGO quasi-autonomous non-government organization
RHA Regional Health Authority
RIBA Royal Institute of British Architects
RICS Royal Institution of Chartered Surveyors
RTPI Royal Town Planning Institute
SIA Shopping Improvement Area
SoS Secretary of State
SPZ Simplified Planning Zone
TCP Town and Country Planning
TCPA Town and Country Planning Association
TWDC Tyne and Wear Development Corporation
UBR Uniform Business Rate
UCO Use Classes Order
UDC Urban Development Corporation
UDP Unitary Development Plan
UP Urban Programme
URA Urban Regeneration Agency
VAT Value Added Tax

PART I
Overview

CHAPTER 1
Basic information and context

1.1 Constitutional and legal framework

Constitutional framework

The United Kingdom is a unitary state consisting of two kingdoms (England and Scotland), one principality (Wales) and one province (Northern Ireland), all of which retain a distinct cultural identity. The Channel Islands (Jersey and Guernsey), the Isle of Man and various Crown colonies, such as Gibraltar and, until 1997, Hong Kong are also associated with the UK, but are not constitutionally integrated with it. Unlike the French overseas *départements* and *territoires*, they are therefore not within the European Community, and so are not included in this book.

The UK Parliament, sitting in Westminster, London, is the only legislature: no other levels of government have any legislative powers. Executive and administrative powers are also highly concentrated in central government hands in London. Parliament is strictly speaking the meeting of two institutions summoned by the monarch – the House of Commons and the House of Lords.

The rôle of the monarch in government is for practical purposes symbolic only, but does include the final formal process of enacting legislation by giving the Royal Assent to Bills passed by Parliament, thus giving them legal force as Acts of Parliament.

The House of Commons is an elected body, with 651 members following the 1992 general election, the previous Parliament having consisted of 650 members (Whitaker 1992). Each member is elected to represent a defined geographical area known as a constituency. To be elected, each member must have received the highest number of votes (i.e. a simple majority, colloquially referred to as "first past the post") at a general election or a by-election in the constituency for which they stood. There is no element of proportional representation in the electoral system and the relationship between votes cast for each party and seats obtained in the House of Commons is highly distorted, especially in respect of minority parties.

Constituencies vary in size and population, and by convention there are assured minimum numbers for Scotland, Wales and Northern Ireland. The distribution is 524 in England, 38 in Wales, 72 in Scotland and 17 in Northern Ireland. In England the average constituency has an electorate of 65,000. Regular reviews of boundaries are undertaken in order to minimize the variation in size of electorate.

The leader of the political party receiving the support of a majority of the members of the Commons (designated MP: Member of Parliament) is appointed prime minister by the monarch. The prime minister in turn appoints government ministers. The most senior ministers (normally 23) form the Cabinet. In constitutional theory the prime minister is first among equals in the Cabinet, and there is collective Cabinet responsibility for the actions of the government. In practice, the power to appoint and dismiss ministers, and to advise the monarch to dissolve Parliament, gives the prime minister the dominant power. Ministers may be appointed from both the Lords and the Commons: the majority are in the Commons but the government must have a set of spokesmen/women in the Lords. In total, approximately 100 MPs are appointed to ministerial posts with responsibility for specified functions of government. A minister is individually responsible to Parliament for his or her assigned Department, and traditionally would be expected to take responsibility for the actions of that department.

Under the Parliament Act 1911 an election of the entire House of Commons is required at least every five years. The present Parliament was elected in April 1992. There must therefore be another general election before April 1997. The exact choice of date is by constitutional convention the choice of the prime minister alone, who decides when to ask the monarch for a dissolution of Parliament. The decision of the prime minister is constrained only if a government loses a vote of no confidence in the House of Commons (this last occurred in 1979) or if the five-year period has elapsed. Following the granting of a dissolution by the monarch, a general election follows within three–four weeks, normally on a Thursday. The leader of the winning party is invited by the monarch to form the next government, and is appointed prime minister, as soon as the outcome of a general election is clear. Transfer of power and the assumption of responsibility then takes place with immediate effect.

The House of Lords is an unelected and continuously existing body with a membership of approximately 800 hereditary Peers of the Realm and approximately 400 life peers appointed by the monarch on the direction of the prime minister, under the Life Peerages Act 1958 (Whitaker 1992). It also comprises 26 of the most senior members of the Church of England and 11 of the most senior members of the judiciary. Attendances can vary dramatically depending on the political agenda and the interest of members.

3

The rôle of the House of Lords is primarily that of a revising chamber, scrutinizing the details of proposed legislation and suggesting amendments.

Legislation must receive the approval of a majority of sitting members in both Houses. The House of Lords cannot reject legislation approved by the Commons but it can delay its enactment for one month in the case of legislation dealing exclusively with finance and two years in any other case. It therefore acts a check on the excesses of any political party enjoying a substantial majority in the Commons, a situation that commonly arises given the electoral system.

The 11 members of the judiciary within the Lords also form the supreme appellate court in the United Kingdom, although in Scotland its jurisdiction extends to civil matters only.

Proposed legislation may be introduced by any Member of Parliament, but is normally proposed by the responsible minister as a Bill. Bills have to be read three times in each House, and undergo scrutiny by a Parliamentary Committee, before they can become law, as Acts of Parliament, by receiving the Royal Assent.

Legislation proposed by government would normally be introduced by the appropriate Cabinet minister or junior minister. The majority of legislation relating to the subject matter of this book would have been the responsibility of the Secretary of State for the Environment or the Secretary of State for Scotland, Wales or Northern Ireland. The former is in effect the English minister but does on occasion introduce legislation on behalf of one of the others as well.

This is normal in relation to Wales, whose legal system is totally integrated with that of England. Scotland, however, still has its own legal system, which unlike that of England is in principle a Roman law system. It is therefore normally necessary to enact separate legislation for Scotland even when the intention is to achieve the same situation as applies in England.

Parliament is in principle sovereign and omnicompetent. It cannot be bound by a decision of an earlier Parliament and its legislation cannot be challenged by UK courts. However, treaty obligations in relation to the European Community do challenge this principle, and this consideration has lain behind the prolonged parliamentary challenge to the Bill to adopt the Maastricht Treaty.

Subnational government

There is no regional government in the UK, although regional administration does exist for Scotland, Wales and Northern Ireland and to a very minimal extent for the English regions.

Local government has no constitutional right to exist. It is created by, and can be abolished by, an Act of Parliament. The rôle of local government is to perform specified functions and services in a manner responsive to local opinion and needs. It is in many ways, however, an executive arm of central government.

The present structure of subnational government is shown in Figure 1. For England and Wales, local government is organized under the Local Government Act 1972. Important changes affecting London and the six major conurbations or metropolitan areas were enacted in the Local Government Act 1985.

Further changes are currently proposed in the form of a rolling pro-gramme. The government has appointed a Local Government Commission charged with the responsibility to review the structure of local government region by region, with the objective of preparing proposals for a single-tier system of subnational government. By mid-1993, proposals had been published for Wales and for a few regions of England. These take the form of abolishing or amalgamating existing counties or districts in order to form larger single-tier units. The government had expressed the aim of implement-ing some of these proposals by April 1994, although in England this is unlikely to be achieved (Municipal Yearbook 1992).

Outside London and the metropolitan areas, there are 39 county councils with an average population of 750,000 and an average area of 310,000 ha. These have delegated primary responsibility within their areas for education, strategic land-use, waste disposal, roads, social services, police, consumer protection and parks (Municipal Yearbook 1992).

Each county is subdivided into a number of district councils (296 in England and 37 in Wales). The average population is 100,000 and average area 40,000 ha. These have primary responsibility for housing, town and country planning, public health, refuse collection, control of commercial premises and urban passenger transport.

The respective responsibilities of the two tiers are not fixed and may vary by agreement having regard to geographic and economic considerations within the two-tier structure.

Within rural districts there are parish councils, but they have only a consultative rôle in local government. Normally this includes the right to be consulted on planning and development proposals.

London and the metropolitan areas (West Midlands, Merseyside, Greater Manchester, West Yorkshire, South Yorkshire and Tyne and Wear) formerly had a similar system of two-tier local government. The 1985 Act abolished the upper tier leaving a single tier with increased powers. Within those areas, compulsory and discretionary joint boards or non-elected councils exist to co-ordinate prescribed functions and services. The principle behind the govern-ment's current proposals is to create single-tier local authorities in the rest of the country. At the time of the general election in April 1992, a change

The UK has a unitary constitution: all legislative power rests with the national Parliament

Level	England	Wales	Scotland	Northern Ireland
National:	Dept of the Environment (DoE)	Welsh Office	Scottish Office Environment Department	Northern Ireland Office DoE (NI)

Regional: Throughout the UK there is no form of elected regional government.

England: No regional administration. Proposals for regional planning guidance

Administered as regions of the UK by departments of central government

Local:

	London	Metropolitan Areas	Non-Met Areas Counties	Counties	Urban Regions	Rural Regions	Island Authorities	Area offices of DoE (NI)
Level I	nil	nil	(39)	(8)	(6)	(3)	(3)	nil
Level II	Boroughs	Metropolitan Boroughs	Districts	Districts	Districts	Districts		Districts
	(32 plus City)	(36)	(296)	(37)	(37)	(16)	nil	(26)

Notes: Local authorities exercising planning powers under the Town and Country Planning Acts are underlined. National Parks authorities, Urban and New Town Development Corporations exercise these powers within their designated territory. Numerals indicate number of authorities in each category.

Figure 1 Government and Planning Authorities in the UK.

of government could have led to proposals for some form of regional authority, as both the Labour and Liberal Democrat parties had such a policy, but the present government has no such intention.

All councils consist of elected members known as councillors. All are elected for a fixed four-year term by an electoral system similar to that for the House of Commons. English and Welsh counties and London boroughs elect the whole council every four years on the first Thursday in May. Metropolitan districts elect one-third in each non-county election year. Non-metropolitan districts in England and Wales have a choice between these two patterns. All Scottish local authorities elect the whole council every fourth year. Districts normally have multi-member constituencies. Councils allocate members to committees and subcommittees for decision-making, important issues being decided or ratified by the full council.

The powers and duties of local government derive from a multitude of legislation, not least the Town & Country Planning Act 1990 and the Planning and Compensation Act 1991. Under section 111 of the Local Government Act 1972 they have a general power to do anything that is designed to be supportive of the discharge of any of their functions, but this power is in effect strictly limited by financial controls exercised by central government.

Local government nevertheless has a degree of independence arising from the manner in which it interprets and performs its statutory functions, which is enhanced by the mandate received from its electorate. It is financed in the following ways:

(a) Local taxes, accounting for approximately 37% of revenue (see Ch.3.3).

Residents pay the Council Tax (which replaced the Community Charge – unofficially known as the Poll Tax – in April 1993); businesses pay the Uniform Business Rate.

(b) Central government grants.

These comprise specific grants for particular functions, plus a general grant called the revenue support grant, and account for approximately 48.4% of revenue.

(c) Borrowing and investment.

(d) Fees, charges and rents (approximately 12.8% of revenue).

(e) Sales of land and buildings, including council house sales (see Ch.1.4).

Local government finances are subject ultimately to the control of central government. High-spending councils can be forced to cut expenditure and reduce services through cuts in central grant. Restrictions can be placed by the Secretary of State on the level of the Council Tax set by any council (a process known as "capping"). The level and purpose of both borrowing and capital expenditure are subject to authorization by the Secretary of State, enabling them to be restricted, which is normally the case.

Local government in Scotland is organized under the Local Government Scotland Act 1973 on similar lines to those applying to the counties in England and Wales, with a two-tier system of regional and district councils and a third tier of community councils in rural districts. Again, reform to create a single-tier structure is intended by the government.

Local government in Northern Ireland is not referred to specifically here as planning responsibility is exercised directly by the Secretary of State for Northern Ireland and the Northern Ireland Department of the Environment.

The powers of the Secretary of State within the planning system, with the discretion it allows in the decision-making process, endow the Secretary of State with considerable personal influence. It could therefore be said that in law the Secretary of State has the final say in the use to which any area of land is put.

For the implementation of particular functions or policies of the state, Parliament commonly delegates powers to bodies and organizations it creates by statute. This is the case, for example, with the Urban Development Corporations set up under the 1980 Local Government Planning and Land Act (see Ch.3.3). Their powers and duties and the level of central government control will be set out in the controlling legislation. A common but unofficial term for such bodies is quango, an acronym for quasi-autonomous non-government organizations, which are nevertheless set up by government to perform some specific task, reporting to the appropriate Secretary of State.

Rights of ownership of land and property

Different legal systems operate within the UK. England and Wales share the same legal system, which is broadly similar to that operating in Northern Ireland, having roots in canon law. The Scottish legal system is distinct, deriving from Roman Dutch law. This section will comment only on English and Welsh law. The relevant features of Scottish legal practice are discussed in the case study from Aberdeen (Ch.6.3).

There is no statutory definition of a plot of land. Plots of land may be freely subdivided and reassembled by reference to any chosen boundaries.

The right of ownership is established by the possession of the interest in land known as the freehold. It involves no relationship with another party, as typifies a feudal system of land tenure. The freeholder is therefore able to occupy land indefinitely without interruption and without a financial or other obligation. The freehold may be freely transferred even when the land (or property) is let under the terms of a lease (see below). The manner of its transfer (known as "conveyance") is however regulated by statute.

The only other method of occupying land to the exclusion of others is under the tenure known as leasehold. Essentially this is a contractual agreement entered into between a freeholder and the leaseholder for the transfer of the freeholder's right of exclusive possession of land for a fixed and determinate period. The contractual agreement or lease creating the relationship of landlord and tenant is binding on their respective successors until the expiry of the agreed duration. The contractual terms regulating the respective obligations of the parties, e.g. rent, repairs, insurance, may be freely negotiated but may in certain instances be overruled by legislation. A breach of the terms of the lease by the tenant gives rise to a claim for compensation on the part of the landlord, and if the breach is "substantial" the landlord may terminate the lease and recover possession. The equity of the termination in all the circumstances of the particular lease may however be challenged in the courts.

In the absence of agreement to the contrary, the leasehold interest may be freely transferred (known as an assignation) or a subsidiary leasehold or sublease created. There is no limit on the number of possible assignations or sublettings. The forms of leasehold that are in most widespread use are discussed below.

Full repairing and insuring leases Such leaseholds dominate the commercial property market and have a normal term of 25 years. Rent is paid quarterly in each year and is reviewed, upwards only, at regular intervals, normally five years. In a multi-occupancy development, a service charge may also be payable if services are provided by the landlord for common parts. Strict controls on the use of the property, assignment, subletting and all aspects of occupation are common. A landlord will seek to protect his investment in the rental income by ensuring that all expenses and costs arising in respect of the property are recoverable from the tenant. On the termination of the lease, the tenant may continue in the tenure under terms set by the courts unless the landlord can demonstrate a real intention to redevelop the property (Landlord and Tenant Act 1954). It is normal and possible for the parties to contract out of this statutory provision.

During 1992–3 it became apparent that many new leases are being created that do not follow the traditional format. They are often shorter and include break clauses that permit the tenant to quit before the end of the lease, and rent changes can go downwards as well as upwards. The conclusion to be drawn is that the "typical" lease in Britain since 1945 has been "maintained" by the strong market position of the landlord vis-à-vis the tenant.

Long leaseholds Land is commonly leased for development by the tenant in all sectors of the property market. Normally for a period of 99 years or longer, long leaseholds are granted on payment of a premium, with only a fixed nominal rent or "ground rent" periodically falling due. Broad restrictions on the form of the development may be imposed, but the contractual terms are normally such as to place the leaseholder in much the same position in his dealings with the land as the freeholder. The land and its buildings in principle revert to the freeholder on the expiry of the leasehold and no compensation is payable. Legislation however enables a leaseholder of a dwelling-house compulsorily to acquire the freehold. Because of the absence of a tradition in building in multi-occupancy units at the time of the development of the principles of property ownership, the law on common ownership is clumsy and ineffective. The position for residential tenancies has been addressed by the Housing and Urban Development Act (1993). This legislation enables willing tenants who hold long leases to buy the freehold of their property from the existing landlord in a form of common ownership. It is anticipated that this form of tenure will be utilized in new developments.

Assured and assured shorthold tenancies The majority of private sector leases of residential property created after 15 January 1989 are regulated by the Housing Act 1988. The main exemptions relate to residential properties whose rateable value exceeds £1500 in Greater London or £750 elsewhere, being therefore "luxury" accommodation and users not deemed to be in need of protection. Other exemptions include properties let at less than two-thirds of their rateable values, i.e. at very low rents, and properties where the landlord is resident. The Act recognizes two tenancies: "assured tenancy" and "assured shorthold tenancy". In respect of the former, a tenant is given security of tenure beyond the original duration of the lease, the landlord being required to show that there are grounds for possession. There are 16 grounds for possession, of which 8 can operate only at the discretion of the courts.

The majority of the grounds relate to fault in the conduct of the tenant, the others being the landlord requiring the premises for occupation by himself or his family and the accommodation being provided in relation to employment that has now terminated. Assured shorthold tenures however predominate, as no security of tenure is afforded to the tenant. They have a minimum duration of six months and on the termination of the lease the landlord has the automatic right to resume possession. A tenant must be given formal notice at the outset of the lease that the lease is of this form, otherwise it will be an assured tenancy. Both forms of tenancy have been subject to rent regulation (discussed in Ch.1.4).

Land assembly Except in the situation referred to below, land assembly for the private sector is a difficult, time-consuming and expensive process. Land ownership is fragmented and there may be a rung of leasehold interests. This is in addition to easements in favour of third parties, which are discussed in Chapter 3.1. All must be acquired piecemeal without the benefit of

10

compulsory purchase powers and the failure to acquire any interest may frustrate the whole exercise. The identification of parties entitled to such interests is also complicated by an incomplete register of property interests (see Ch.3.1).

Rights of compulsory purchase are available to the private sector however if sought under a private Act of Parliament. A private Act follows the same procedure through Parliament as a public Act. In practice, private Acts have been successful only for major works relating to infrastructure. An important recent example is the private sector's rôle in the finance and construction of the Channel Tunnel where powers of compulsory purchase were given under the Channel Tunnel Act 1987.

Public authorities almost invariably enjoy powers of compulsory purchase, but these powers may be exercised only in the performance of their statutory functions. Urban Development Corporations (UDCs), local authorities and other economic development agencies do however play an important rôle in assembling land for development in partnership with the private sector. The right of compulsory purchase will be contained in the particular legislation establishing the public authority. The procedure is laid down as a common code in the Acquisition of Land Act 1981. Briefly this is as follows. The authority makes a Compulsory Purchase Order and notifies all interests in the land in addition to the general public. Objections may be made and, if sustained, a public inquiry is held. The order is then confirmed or rejected by the government minister responsible for the authority. If it is confirmed, the authority serves a notice to treat under the Compulsory Purchase Act 1965, which invites parties holding an interest in the land to make a claim for compensation. Compensation is negotiated and the District Valuer normally assists agreement, but, failing agreement, the compensation is settled by the Lands Tribunal.

An individual is in principle free to bequeath ownership of property or his interest under a leasehold tenure to anyone he chooses. There are certain exceptions as regards leases, the major ones being that a joint tenant automatically succeeds the deceased to the full tenancy, and that a landlord of an assured tenancy is entitled to resume possession within 12 months if the successor is a party other than the spouse of the deceased. A dependent relative for whom no provision is made in the will may obtain a settlement through the courts, but a property settlement will be made only in exceptional circumstances. If, however, no will is made then the estate of the deceased is distributed in accordance with the terms of the Administration of Estates Act 1925. In the Act no distinction is made between land and moveable property, and in the absence of agreement between the members of the classes of relatives entitled to succeed there is the presumption that all assets will be realized for distribution in accordance with the terms of the Act.

Table 1.1 National economic statistics, 1980–92.

	GDP	B/P	S	U	W	Pr	GS	I
1980		3113	23.4	6.8	26.8	88.3	(10,705)	100
1981		6226	22.5	10.5	26.7	89.7	(7,265)	112
1982	280	4032	22.5	12.2	26.7	92.5	(7,515)	121
1983	290	3163	21.3	12.7	26.6	97.0	(9,567)	127
1984	296	1562	23.2	11.7	27.3	98.0	(10,177)	133
1985	308	3763	22.5	11.8	27.8	100.0	(8,093)	141
1986	319	(66)	18.7	11.8	27.9	103.7	(7,953)	146
1987	334	(4322)	19.7	10.6	28.0	106.6	(5,242)	152
1988	349	(16,179)	17.8	8.4	28.3	107.9	4,844	160
1989	357	(21,726)	23.5	6.3	28.4	107.5	4,752	172
1990	359	(17,029)	31.7	5.8	28.5	107.5	(7,077)	191
1991	350	(6,382)	39.7	8.1	28.3	107.8	(16,398)	202
1992	348	(11,916)	50.7	10.1	28.0	110.3	(37,365)	138

Sources: Financial Statistics, National Income Accounts, Monthly Digest of Statistics.
Notes: GDP = gross domestic product (3bn), B/P = balance of payments (£m), s = savings (£bn), U = unemployment (%), W = workforce (m), Pr = labour productivity index, GS = government surplus (£ million), I = index of retail prices.

1.2 The economic framework

Data on economic development since 1980

Much of the information provided here takes the form of summary data presented in tabular form (Tables 1.1–1.4). It should be noted that there is an absence of explicit data on leisure development, although it is reflected in Sectors 5 and 10 in Table 1.2. The leisure industry is nevertheless a significant growth sector in the economy. The figures for gross national product (GNP) and GNP per head indicate that for much of the period Britain has achieved rates of real economic growth that are quite comparable with the performance of other industrialized countries. At the beginning of the 1980s a short-term recession brought growth to a halt and since early 1990 the economy has moved into a more sustained depression. This cyclical pattern is not unusual in Britain.

The sectoral analysis of the economy shows quite clearly that the growth in output of the sectors of the economy has not been uniform. Particularly rapid growth has occurred in distribution, banking, education and health, and other services. By comparison, the manufacturing sector has shown little real

Table 1.2 Gross domestic output by sectors, 1981–91 (at current prices, £ billion).

Year	Sectors									
	1	2	3	4	5	6	7	8	9	10
1981	4.8	23.5	54.8	13.0	27.3	15.9	13.9	16.2	20.6	12.0
1982	5.5	26.1	59.5	14.1	30.1	17.2	15.0	17.4	21.3	13.3
1983	5.3	30.1	62.1	15.7	33.2	18.3	15.9	18.8	23.4	15.1
1984	6.5	29.7	66.4	16.9	36.2	19.8	16.7	20.2	24.6	16.6
1985	5.9	33.0	72.6	17.7	40.5	21.2	18.0	21.4	26.6	18.4
1986	6.6	24.4	77.8	20.1	44.8	23.0	19.6	23.0	30.0	20.7
1987	6.9	25.3	82.6	23.1	48.9	24.8	21.1	24.9	33.2	23.3
1988	7.0	23.1	91.2	27.0	56.6	27.6	23.3	27.5	37.6	26.2
1989	8.1	23.8	97.4	30.3	62.1	30.1	25.5	29.6	42.5	29.7
1990	8.7	25.5								
1991	8.8	28.3								

Source: National Income accounts.
Key to sectors: 1, Agriculture, Forestry and Fishing; 2, Energy and Water Supply; 3, Manufacturing; 4, Construction; 5, Distribution, Hotels and Catering; 6, Transport and Communication; 7, Banking, Finance and Business Services; 8, Public Administration and National Defence; 9, Education and Health; 10, Other Services.

Table 1.3 Net domestic fixed capital formation by industry group, 1981–91 (at current prices, £ million).

Year	1	2	3	4	5	6	7	8	9	10	11
1981	(164)	1,047	374	(835)	(129)	1,349	(852)	2,752	2,584	3,537	9,663
1982	56	1,004	285	(1,069)	(54)	1,581	(1,133)	3,034	3,362	4,105	11,171
1983	135	730	399	(1,176)	(24)	1,595	(901)	2,900	3,542	5,265	12,465
1984	133	843	(269)	44	(95)	2,307	(265)	3,443	4,150	5,918	16,209
1985	(177)	354	(122)	1,263	(72)	2,842	165	4,183	4,566	5,486	18,488
1986	(198)	(21	(30)	453	(111)	3,067	(382)	4,857	5,118	6,676	19,430
1987	(167)	(611)	(213)	1,015	7	4,189	962	7,699	5,457	7,591	25,928
1988	(51)	(559)	(80)	2,192	355	5,690	2,288	10,756	6,115	10,554	37,261
1989	(93)	(268)	155	3,443	261	5,301	3,399	15,093	8,610	10,729	46,630
1990	(261)	461	699	2,709	55	4,429	2,482	15,444	10,391	8,492	44,902
1991	(393)	2,022	194	801	na	na	na	na	9,506	4,678	31,474

Source: National Income Accounts.
Key to industry group: 1, agriculture, forestry, fishing; 2, extraction of oil and gas; 3, other energy and water supply; 4, manufacturing; 5, construction; 6, distribution, hotels, catering; 7, transport & communication; 8, banking, finance, insurance, business services; 9, other services; 10, dwellings; 11, total.
Notes: The figures are gross domestic fixed capital formation less capital consumption.
Assets are classified to industry on basis of ownership and not use.
na = not available

Table 1.4 Quarterly interest rates, 1980–1993.

		3MIBR	MR			3MIBR	MR
1980	Q1	17.68	14.0	1987	Q1	10.64	12.3
	Q2	17.15	14.0		Q2	9.16	11.7
	Q3	16.12	14.0		Q3	9.80	11.3
	Q4	15.53	14.0		Q4	9.22	11.0
1981	Q1	13.32	14.0	1988	Q1	9.00	10.3
	Q2	12.48	13.5		Q2	8.41	10.0
	Q3	14.25	16.6		Q3	11.34	11.0
	Q4	15.62	15.0		Q4	12.49	12.7
1982	Q1	14.36	15.0	1989	Q1	13.06	13.3
	Q2	13.36	13.5		Q2	13.49	13.5
	Q3	11.46	13.0		Q3	13.93	13.5
	Q4	9.93	11.0		Q4	15.08	13.5
1983	Q1	11.18	10.0	1990	Q1	15.19	14.8
	Q2	10.16	10.0		Q2	15.11	15.3
	Q3	9.83	11.25		Q3	14.96	15.2
	Q4	9.37	11.25		Q4	13.83	14.8
1984	Q1	9.27	11.0	1991	Q1	13.23	14.3
	Q2	9.25	11.0		Q2	11.58	13.2
	Q3	11.14	12.5		Q3	10.77	12.0
	Q4	10.14	13.0		Q4	10.44	11.4
1985	Q1	13.02	12.8	1992	Q1	10.55	11.2
	Q2	12.62	14.2		Q2	10.21	10.9
	Q3	11.69	13.8		Q3	10.16	10.6
	Q4	11.62	13.0		Q4	7.58	9.9
1986	Q1	12.41	13.0	1993	Q1	6.36	8.4
	Q2	10.18	12.1				
	Q3	9.99	11.0				
	Q4	11.21	11.8				

Source: Financial Statistics.
Key: 3MIBR = the three month inter-bank rate, MR = the mortgage rate.

growth. The figures generally reflect the increasing de-industrialization of Britain and the movement towards a service-centred economy. The latest figures show clearly the severity of the depression in retailing and construction.

The figures for the investment rate per sector point to both the wide oscillations in the British economy during the 1980s and the growing importance of the service sector. The manufacturing slump at the beginning of the period is quite apparent, as is its later recovery. The data reflect the changes in consumer demands and are consistent with changing employment patterns.

The levels of investment, which include investment in buildings, have implications for the form and pattern of demand for premises. For example, the net disinvestment in manufacturing during the 1980s was associated with a decline in the demand for factory premises. Also the growth in service sector employment has meant a growth in the demand for office space.

The unemployment, inflation and balance of payments figures show some well known characteristics of the British economy. Unemployment has been much higher than during the 1950–75 period, and in the 1992–3 period reached almost 3 million. Inflation continues to be a problem, with the price level almost doubling between 1980 and 1990, though by June 1993 it had fallen to 1.3%. Wage inflation has generally been above price inflation and well above the rate of growth of productivity.

The balance of payments in Britain has tended to move into surplus during periods of economic depression but into deficit as the economy expands. The evidence of the 1980s is that this process continues, though through the recession of 1990–3 the balance of payments has remained in substantial deficit.

The labour force figures show the continued growth in potential employees. The main categories of growth are part-time and women workers. This is consistent with the growth of a low-paid service sector. Including part-time workers, the labour force is still growing and is at an all-time high.

The rate of interest in Britain is determined by the money markets, and government influence is exerted through its borrowing stance. During the 1980s the government used monetary policy to restrict the rate of growth of demand. This was difficult for much of the period, resulting in interest rates that were higher than those in many other countries. From the beginning of 1988 the government sought to reduce the rate of growth of borrowing by successive increases in rates of interest.

With the onset of the economic recession in 1990 there was a need to reduce interest rates rapidly. While some reduction occurred, low interest rates were seen as incompatible with Britain's membership of the Exchange Rate Mechanism (ERM) at the then current rate. Despite the government's attempts to control the exchange rate, market pressure forced the pound sterling out of the ERM in October 1992. Since then there has been a dramatic fall in interest rates and the value of the pound sterling. This policy change has been positively endorsed by much of industry and commerce.

The issue of capital mobility is not pertinent in Britain. Exchange controls were abolished at the beginning of the 1980s and capital has flowed out of the country in every subsequent year. Many of these investments have been highly profitable and they now generate a significant inflow of funds to Britain. The abolition of controls may be one of the reasons why Britain has been an attractive home for international business.

1.3 The social framework

Demographic development

The 1991 Census figures show the UK population was 55,500,000. The population of its constituent elements was: England and Wales – 48,960,000; Scotland – 4,957,000; and Northern Ireland – 1,583,000.

Table 1.5 provides information on population changes and projections and the influence of births, deaths and migrations. Between 1971 and 1989 the population of the UK showed a slow rate of growth – only 1.3 million or 2.3%. This is largely attributable to the collapse in the birth rate from the baby boom experienced between 1959 and 1964, with births peaking in 1964 at 1,015,000, to a decline in births through the 1970s to a trough of 657,000 in 1977. The birth rate has since recovered and is estimated to have peaked in 1990 at 830,000 as the large generation born in the early 1960s reached its child-bearing years.

The baby boom of the 1960s and the survival, through advances in modern medicine, of the large numbers born in the first 30 years of this century have

Table 1.5 Population changes and projections, 1901–2011 (thousands).

		Average annual change					
	pop^n at start of period	live births	deaths	net natural change	net civil migration	other. adjust- ment	overall annual change
Census figures							
1901–11	38,237	1,091	624	467	(82)		385
1911–21	42,082	975	689	286	(92)		194
1921–31	44,027	824	555	268	(67)		201
1931–51	46,038	785	598	188	25		213
Mid-year estimates							
1951–61	50,290	839	593	246	(9)	15	252
1961–71	52,807	963	639	324	(32)	20	312
1971–81	55,928	736	666	69	(44)	17	42
1981–88	56,352	745	657	88	10	4	102
1988–89	57,065	778	639	139	38	(6)	171
Projections							
1991–96	57,333	835	649	186			186
1996–01	58,462	801	654	148			148
2001–06	59,201	743	653	89			89
2006–11	59,648	725	657	68			68

Source: Social Trends 1991.

had a significant effect on the age structure of the population (see Table 1.6). A closer analysis is undertaken for 1971–89. The fluctuation in the birth rate and the changing age structure of the population have very wide-ranging implications, not least in the land and property markets. There has been a significant decline in the school-age population, mirroring a comparable increase in the young adult population, the most active group in household formation. Also of significance is the growth in numbers of those over pension age and in particular those over 80.

Table 1.6 Age structure of population, 1961–2006 (millions).

	under 16	16–39	age group 40–64	65–79	80+	all ages
mid-year estimates:						
1961	13.1	16.6	16.9	5.2	1.0	52.8
1971	14.3	17.5	16.7	6.1	1.3	55.9
1981	12.5	19.7	15.7	6.9	1.6	56.4
1986	11.7	20.6	15.8	6.8	1.8	56.8
1989	11.5	20.4	16.3	6.9	2.1	57.2
1991	11.7	20.2	16.5	6.9	2.2	57.5
mid-year projections:						
1996	12.5	19.8	17.0	6.8	2.4	58.5
2001	12.8	19.2	18.0	6.7	2.5	59.2
2006	12.1	18.1	20.2	7.0	2.7	60.0

age group	1971 '000	%	1989 '000	%	1971–89 change '000	%
under 16	14.3	25.6	11.5	20.1	(2.8)	(19.6)
16–39	17.5	31.3	20.4	35.7	2.9	16.6
40–64	16.7	29.9	16.3	28.5	(0.4)	(2.3)
65–79	6.1	10.9	6.9	12.1	0.8	13.1
80+	1.3	2.3	2.1	3.7	0.8	61.5

Source: Social Trends 1993.

Over the last 30 years there has been a significant increase in the numbers of people living alone (see Table 1.7). In 1989, a quarter of all households on mainland UK contained only one person, compared with one-eighth in 1961. Between 1961 and 1989 the proportion of households containing five or more persons halved from 16% to 8%. These changes contributed to a reduction in the average household size from 3.09 people in 1961 to 2.51 people in 1989. The growth in the single-person household is attributable to the growth in the elderly population, the rise in the level of divorce and the increasing tendency for young adults to live alone (see also Table 1.10).

Table 1.7 Households by type and size in England, Wales and Scotland, 1961 –91, (%).

	1961	1971	1981	1991
Households by type:				
1 person households				
under pensionable age	4	6	8	11
over pensionable age	7	12	14	16
2 or more unrelated adults	5	4	5	3
1 family households:				
no children	26	27	26	28
1–2 dependent children	30	26	25	20
3 or more dependent children	8	9	6	5
non dependent children	10	8	8	8
lone parent with:				
dependent children	2	3	5	6
non dependent children	4	4	4	4
2 or more families	3	1	1	1
all households	100	100	100	100
Households by size:				
1 person	12	18	22	26
2 people	30	32	32	34
3 people	23	19	17	17
4 people	19	17	18	16
5 people	9	8	7	6
6 or more people	7	6	42	2
all households	100	100	100	100
average household size (no. of people)	3.09	2.89	2.71	2.48

Source: Social Trends 1993.

The UK relationship with the Commonwealth has seen a significant exchange of populations over the last 40 years as white UK citizens emigrated to Canada, Australia, and New Zealand to be replaced by ethnic groups from

Table 1.8 Population by ethnic group and age, 1989–91, (% and '000).

Age	West Indian & Guyanese	Indian	Pakistani	Bangladeshi	Chinese	Arab	African	mixed	other	White	not stated	all groups
					Ethnic group							
0–15	25	31	44	49	29	22	27	54	28	20	28	20
16–29	32	27	25	22	26	32	31	26	27	22	24	22
30–44	16	24	18	14	31	29	26	11	28	21	18	21
45–59	20	14	11	13	9	11	12	6	11	17	13	16
60–	7	5	2	2	5	6	3	3	6	21	18	21
000s	456	793	486	127	137	67	150	310	155	51,805	496	54,983

Source: Labour Force Survey, Office of Population Census and Surveys.

the New Commonwealth, predominantly from India, Pakistan and the Caribbean. Immigrants granted the right to live and work in the UK are also entitled to receive British citizenship and a UK (now EC) passport. In 1951 the non-white ethnic minority community was estimated at 200,000. It grew significantly to an estimated 2.5 million in 1989 (see Table 1.8).

Mobility

Three key dimensions have characterized geographical mobility of the population during the post-war period: (a) the change in the balance between north and south; (b) the urban–rural shift from larger metropolitan centres to smaller cities and towns, including those situated in relatively remote areas; (c) suburbanization and city–regional decentralization from the larger metropolitan centres to their commuting hinterlands. The first of these became more dominant during the 1980s. Previously the other two were more significant. The 1991 Census indicates continuing interregional redistribution of the population, as Table 1.9 shows. More recent data however show some reversal of this trend.

Table 1.9 Population changes by region, 1981–91.

Region	% change
Scotland	(3.4)
North	(3.0)
Yorks/Humberside	(1.8)
North West	(4.3)
Wales	0.3
West Midlands	(1.4)
East Midlands	2.5
East Anglia	7.7
South West	5.5
Outer South East	3.0
Greater London	(4.8)

Source: 1991 Census Report.

Continuing movement away from the metropolitan areas to smaller towns and self-standing cities and to semi-rural and retirement areas is confirmed by the preliminary results. All major conurbations showed population decline, the biggest declines being Merseyside (–9.0%), urban Strathclyde (–7.7%), inner London (–5.9%) and Birmingham (–5.5%). The biggest single increase was in the district containing the new town of Milton Keynes, with a growth of 39%.

Changes in social values

The election of the Conservative government in Britain in 1979 has been widely identified as the break-point between the post-war political consensus and the more competitive and conflictual 1980s. The social trends identified reflect the most obvious changes in values and practices during the 1980s.

Privatization During the 1980s a large proportion of the industries owned by the public sector, or nationalized industries, were returned to the private sector. This economic policy was widely accepted because of the belief that the private sector is more efficient and more responsive to consumers. This is now generally accepted for industries operating in a competitive environment, such as British Steel, British Airways and even British Telecom, but remains controversial in the case of services that are both essential for life and deemed to be "natural monopolies" such as the water supply industry. In excess of 10 million people now own shares in privatized companies.

Unemployment The level of unemployment during the 1980s was much higher than at any time since the 1930s. The avoidance of unemployment was seen in the 1960s as an essential element of government policy, but by the mid-1980s the main objective of economic policy was seen as control of inflation. Unemployment has been seen as in part an inevitable consequence of the restructuring of the economy, in part the consequence of trade unions pushing up wages and in part the fault of the unemployed themselves who by a lack of skills and aptitude have made themselves unemployable.

Divorce The level of divorce began to grow after the reform of the divorce laws in the 1960s. During the 1980s divorce was made easier and quicker and the level of divorce has risen until today it is estimated that 40% of all marriages will end within 10 years. In over 80% of the divorces where children from the marriage are involved the custody of them is given primarily to the mother.

One-parent families Given the high levels of divorce, a growth in the number of one-parent families is inevitable. The actual growth has been compounded by the number of single women who have children. Some of the women have permanent partners and are therefore considered to live in conventional households without the "benefit" of the marriage contract. It is estimated that approximately 2 million one-parent households exist (over 90% of them headed by females), with the result that many children grow up without a father in their house for a large part of their childhood. For many of these households the problem of poverty is both serious and persistent.

One-person households The most numerous category of household size is now one-person households. This trend has been magnified by the ageing of the population. This has created a need for additional housing units, although many of the households cannot afford housing without financial support from the state (see Ch.1.3 and Table 1.10).

Table 1.10 Actual and projected numbers of households, England and Wales, 1981–2001, ('000).

Year	One parent families	One person households	All households	Population
1981	1,479 [100]	4,126 [100]	18,338 [100]	48,918 [100]
1983	1,581 [106.8]	4,352 [105.5]	18,444 [100.6]	48,937 [100]
1985	1,692 [114.4]	4,622 [112.0]	19,016 [103.7]	49,207 [100.6]
1987	1,800 [121.7]	4,875 [118.2]	19,422 [105.9]	49,527 [101.2]
1991	1,976 [133.6]	5,386 [130.5]	19,989 [109.0]	50,009 [102.2]
1996P	2,127 [143.8]	6,012 [145.7]	20,735 [113.1]	50,822 [103.9]
2001P	2,189 [148.0]	6,533 [158.3]	21,215 [115.7]	51,455 [105.2]

Source: Housing and Construction Statistics.
Notes: P = projection

Homelessness Homelessness has always existed in Britain amongst people commonly seen as tramps and drunks. During the 1980s the numbers of homeless rose and the types of people changed.

Numbers are difficult to estimate, for these people are rarely surveyed or recorded. The largest concentrations exist in London, where over 40,000 are estimated to be homeless. It is, however, a social problem in every city and rural area.

The new homeless come primarily from two sources. Young people, especially young men, make up one of the fastest-growing groups. They have little access to public housing and have often become homeless through a breakdown in their family circumstances associated with parental divorce or poverty.

An additional source of homeless people comes as a result of the discharging of patients from mental hospitals under the government's new social policy known as "care in the community". Despite much effort, many of these people end up homeless and they can become a serious danger to themselves and others.

Residualization This term has been most widely used in the housing field to describe the rôle of public housing in the 1980s. With the sale of good-quality public housing to the tenants and the increasing preference for owner-occupation amongst all sectors of the population, the state has appeared to say that public housing is only for the residual who are unable to obtain housing in any other way.

More widely the term has been used to identify those sectors of the population who are seen to be outside the central stream of society and in particular those whose standard of living does not permit them to enjoy the goods and services that form a normal part of most people's lives. Such people are typically unemployed, low skilled and/or part of a one-parent family. They typically occupy poor-quality housing and suffer from low educational achievement and poor health. An alternative method of identifying this group is to say that they suffer from multiple deprivation.

Changes in work patterns One of the most significant economic changes in the 1980s in Britain was the growth in non-traditional work patterns. A lot of part-time and job-share posts and temporary work were created, while the availability of permanent full-time employment declined. Even where full-time work remains, the substantial restructuring of the economy has resulted in changes in work skills, with the result that far fewer people can assume that their employment is really permanent.

Another change was the growth in female employment. It is now normal to expect women to work both before childbirth and soon after it. Many of the jobs that they take, especially after starting a family, are of a part-time nature and they are often low paid.

A further change was the growth in service sector employment as Britain moved towards a post-industrial economy. Already over 50% of the working population are employed in work defined as non-manual and all of the growth in employment is in this sector.

Overall, Britain has moved from an economy dominated by male, manual, full-time employment to one where the work patterns are far more diverse. Partly this reflects changes in production technology and partly it reflects changes in spending patterns as an increasingly rich society chooses a wider range of goods and (most importantly) services.

1.4 Land and property markets, and the building industry

Commercial property

The commercial property market in Britain can be considered as the sum of three separate markets: retail, offices and industrial property. One common characteristic of each sector is the division of the market into an occupiers' market and an investment market. These are quite separate from one another but not unconnected. The division is created because of the widespread practice of occupiers renting their property requirements from a landlord by signing a lease, the details of which are explained in Chapter 1.1.

In the occupier market, the level of rents is determined by demand and supply. The office and retail sectors are unaffected by government policy (except in Enterprise Zones and Urban Development Corporation areas), but in areas of industrial decline, such as the north of England, government agencies (e.g. English Estates) affect supply and therefore price.

In the investment market, the value of buildings is determined by the demand for and supply of buildings as investments. Buyers purchase them in anticipation of future rental income, but great uncertainty occurs here, especially when the building is not let. The existence of a tenant and the obligations of the lease bring greater certainty to the owner and enable property to be seen as an attractive investment. The case study of the London office market (Ch.9.1) provides an example of the functioning of occupier and investment markets.

Retail sector The retail sector is spatially dispersed throughout the country, with high rents and capital values being achieved in a variety of locations. Modern city centre retail developments and out-of-town developments are commonly undertaken by a property developer and then rented to occupiers, who also pay a service charge for the management of the whole centre by a specialist management team.

Office sector This sector is concentrated in urban centres and most particularly in London. The predominant tenure is leasehold, with rents and capital values determined by market forces. The most popular new form of development is the business park.

Industrial sector This sector should be considered as split into specialized buildings such as oil refineries and steel plants and general-purpose industrial buildings. The former should be treated as capital investment in plant and such structures are commonly owned by the user. No real market for these buildings exists, though they do create land requirements and they do have an influence on the value of adjacent land.

General-purpose industrial buildings are those that can be leased to a variety of users, though manufacturing industry and warehousing dominate. The historical tradition up until 1939 was for occupiers to purchase their own premises, and many larger well established companies still hold considerable freehold property units. Since 1945 there has been a rapid growth in the leasehold tenure, until today it dominates the functioning of the market. The explanations for this change are complex but two issues are commonly cited: that firms can more profitably use their scarce and expensive capital in ways other than property purchase; and that for small firms it is far easier to begin business with rented property than via property purchase.

Finally it should be note that many industrial properties are nowadays almost indistinguishable from offices, as the case study of Stockley Park (Ch.6.1) illustrates.

Owner-occupation of commercial property   No detailed figures are available for non-residential owner-occupation.

It is widely believed that owner-occupation is irrelevant in the office sector. In retailing, owner-occupation exists but most of the newer developments both within the city centre and on peripheral sites are let to tenant occupiers. In the industrial sector, owner-occupation is important and possibly as much as 40% of the space is held in this way. The market is however dominated by the leasehold tenure.

Changes of ownership   Change of ownership of property is made possible in Britain by the existence of a widespread and transparent property investment market. Changes may be from investor to investor, from owner-occupier to investor, from investor to owner-occupier and from owner-occupier to owner-occupier. In addition, an owner-occupier may sell a property to an investor and simultaneously rent the same property back on a lease.

In principle it is always possible to buy or sell a property, the only condition being an acceptable price. In practice the statement is true for a large range of property but it may be very difficult to find a buyer for poor-quality property in depressed regions, especially if it is not occupied by a tenant.

The investment market tends to set the price levels and these vary with local conditions of demand and supply. Information is provided locally by firms of chartered surveyors, members of whom are almost always involved in commercial property transactions.

For much of the post-1945 era the investment market has been dominated by the financial institutions. Their definition of "acceptable investment property" (based on location, type, quality, etc.) has tended to force developers to build to investors' specifications rather than to meet the needs of occupiers.

Changes of ownership tend to be tied to the property cycle, with the number of changes peaking just before prices reach their highest level.

Residential property

Historical background During the 20th century considerable changes have taken place in the housing sector, with the result that the current state of the sector is most unlike that existing in the pre-1914 period. The most significant change has occurred in the form of housing tenure. While before 1914 over 80% of the population lived in accommodation rented from private landlords, by 1981 this had fallen to 11.1%. The growing sectors have been owner-occupation, reaching 29% of households by 1950 and 68% by 1991, and local authority renting, achieving 18% in 1950 and 31% in 1975 (see Tables 1.11 & 1.12, and Balchin 1989).

Table 1.11 Housing tenure in 1990 by region, (%).

Region	Owner occupation	Local authority	Private renting	Housing association
North	60	29	7	4
Yorks/H'side	66	24	8	2
E. Midlands	71	20	8	2
E. Anglia	70	17	10	2
S. East	69	19	8	4
S. West	74	15	10	2
W. Midlands	68	24	6	3
N. West	68	22	6	4
England	69	21	8	3
Wales	72	19	7	2
Scotland	51	40	6	3
N. Ireland	66	29	4	2

Source: Regional Trends 1990.

Owner-occupation tenure growth derives both from changes in the tenure of existing buildings and from a process of building new houses for sale by private builders. Owner-occupation has become the tenure preferred by a substantial majority of the population and its growth has been encouraged by successive governments by means of tax benefits, land supply mechanisms and on occasion direct subsidy. Additionally the tenure has become more financially attractive with the appearance of prolonged periods of inflation in Britain.

Table 1.12 Changes in housing tenure in the UK, 1914–91, (%).

Year	Owner occupation	Local authority	Private renting & housing association
1914	10	<1	90
1950	29	18	53
1960	42	26	32
1975	53	31	16
1985	62	27	11
1989	66	23	10
1991	68	22	10

Source: Housing and Construction Statistics, 1979–93.

Provision of subsidized social housing by local authorities (known as council housing) became common in the 1920s. The sector grew substantially in the post-1945 period through an active policy of compulsory purchase and demolition of housing deemed unfit for human habitation (slum clearance) and the utilization of the cleared land for the provision of housing for those groups who were unable to afford entry into the owner-occupied sector (see Table 1.13 below). The council housing sector at its peak in the late 1970s accounted for over 30% of the national housing stock, and over 50% in certain cities.

Another key sector of public authority rented housing from the 1950s to the 1980s was that of new town housing, built and let by the New Town Development Corporations. The majority of these dwellings are now owner-occupied, and the remainder, predominantly special housing, is now in local authority or housing association hands.

Private renting declined partly because households preferred other tenures but also because other tenures were subsidized and supported by the state while the privately rented sector faced rent control and penal taxation of rent income (Hamnett & Randolph 1988). Private renting nevertheless is recognized by governments as important both in the maintenance of a means of access to housing for new households and also as a means of facilitating the mobility of the working population.

The changing pattern of tenure and the means by which these tenures have been created have had a major impact on the urban form of Britain's cities. Local authorities have been able and willing to develop within the existing urban sphere partly because each local authority had a rôle in housing its local electorate within its local boundaries and partly because they could use compulsory purchase powers to assemble sites, using their housing as well as their planning powers. As a consequence, large inner urban areas are now dominated by council dwellings, or recognizable former council dwellings with their very distinctive street patterns and architectural forms, discernable in spite of extensive rehabilitation, as the Falstone Walk case study (Ch.9.3) illustrates.

Private developers of housing for owner-occupation have found development within urban areas more difficult owing to problems of land assembly, externalities and the higher construction costs associated with redevelopment of urban land. They have preferred to develop on "greenfield" sites and consequently have been a leading pressure group in the campaign for a greater release of land for building on the outskirts of urban areas. Over time, urban areas have expanded and much of the perimeter is now used for relatively low-density private housing of the type illustrated in the Hertfordshire and Aberdeen case studies (Ch.6.3 and Ch.9.2).

This urban form with owner-occupation at the edge and council accommodation at the core predominates, but there are many examples both of private housing development within the core and local authority housing on the periphery. The main pressures for urban spatial expansion have come, however, from the private sector.

Demand for owner-occupied housing The demand for owner-occupation is a function of both demographic trends and the financial circumstances of the purchasers. While today roughly half of all owner-occupiers own their property outright, the market demand is driven by the price and availability of mortgage finance and the existence of subsidies.

Borrowing to purchase a house in the early working years of one's life is considered normal and a range of financial institutions have evolved to facilitate the process. Building societies are the main and traditional source of home loans or mortgages (indeed this is their historic purpose), but clearing banks also offer this facility. At times of high UK interest rates, non-sterling mortgages (especially in Swiss francs and Deutschmarks) have been attractive to some people seeking large loans (see Ch.9.2).

During the 1980s it was possible to borrow a sum roughly equivalent to three times the household's annual earnings and up to 100% of the property's value. Repayments were made monthly for a period up to 25 years. With the fall in house prices during 1990-3, many householders now are in the situation of having negative equity. In other words, their total of outstanding mortgage debt, which may be the result of a 90% or 100% mortgage, is now greater than the market value of their house. As a consequence of this phenomenon, lenders have become less willing to lend up to the limit, especially on housing in less prosperous areas.

It is normal practice for borrowers to move house and use the increase in property prices and increased borrowing as a means of purchasing a more

29

expensive property. This situation has been facilitated by the continuing process of inflation in Britain. Borrowing is curtailed only by the borrower's financial standing, interest rates and house valuations. Borrowings are normally made on terms that include flexible interest rates.

Supply of owner-occupied housing The supply of owner-occupied housing is inevitably price inelastic in the short run and consequently price determination depends solely on the level of demand (see Table 1.13).

Table 1.13 Housing completions by sector in England and Wales, 1976–91.

Year	Private enterprise[1]	Housing associations[2]	Local authorities[3]
1976	138,477	14,618	124,512
1981	103,902	17,378	58,147
1986	154,968	10,998	20,258
1988	179,303	10,666	16,668
1989	157,215	10,809	13,555
1991	132,291	17,603	8,569

Source: DoE.
Notes: 1, These were almost solely for owner occupation; 2, These were primarily for renting though some sales occurred; 3, Includes New Towns. These were for renting though sales under the "right to buy" legislation may have occurred.

The supply of land for house building has been a contentious issue for much of the post-war period. In 1981, the current government decided that each local authority should ensure, as part of its planning policy, that sufficient land would be made available to ensure that a five-year supply was continuously available. This land had to be free from ownership and planning constraints, suitable for development and capable of meeting the market-determined needs of house purchasers. This requirement is no longer enforced but local plans are expected to designate sufficient land for housing.

Housing development in the owner-occupied sector is undertaken by private builders, who often make considerable profits from the purchase of land in advance of planning permission being available. This "land banking" (see Ch.3.2) is widespread and is in part the consequence of a discretionary planning system.

Regional house price differences Considerable differences in the price of comparable housing exist between regions, and as each regional average

30

includes wide variations there is even greater disparity at local levels. Over time, regional differences have been at least maintained. They grew very wide at the height of the late-1980s boom. These differentials have had two major consequences. First, those who have seen the real price of their house rise fastest have benefited from real increases in their wealth, while prospective new entrants to the housing market have been seriously disadvantaged. Secondly, geographical mobility in search of employment opportunities has been discouraged, especially in the case of migration to the South East of England where jobs were more readily available during the 1980s.

During the 1990–3 period, regional differences narrowed as house prices (and, in some sectors, job prospects) in the South East fell more than elsewhere.

Wealth effects The high proportion of owner-occupiers in the housing sector has, over time and especially in the inflationary period since 1945, led to a build-up of housing wealth. This wealth tends to be released at certain periods. First it will be released when the house-owner dies and passes the estate to their heirs. Considerable capital sums are then available to people who often already own their own house.

Secondly, it is common when purchasing for the second or subsequent time to borrow an additional sum in excess of the difference between the two house prices. Although such withdrawal of capital may be necessary to finance new carpets, curtains, etc., the practice of withdrawing large sums for bigger purchases is common. When the housing sector is very active it is quite common for very large sums, amounting to several billion pounds nationally, to be withdrawn and used to finance general consumption, stimulating the retail sector with consequent property market effects.

Alternatives to owner-occupation of housing
Private renting The prolonged period of decline in this sector has meant that it now cannot perform the function of the "normal" tenure. Roughly half of the tenants are elderly people who have been in their homes for many years, their tenancy protected by legislation. The remaining tenants are typically younger, more mobile and often restricted in some way from access to council housing. Prior to 1980, the tenancies were difficult to arrange within the law, although many did exist in forms that avoided the tenancy regulations.

During the 1980s the current government attempted to support the tenancy by the creation of shorthold tenancies (see Ch.1.1) where the tenant has security of tenure for a short fixed period and where, today, the rents are freely determined by landlord and tenant. Most tenancies are now created within the terms of this legislation.

The government has also attempted to encourage investment in property for renting by schemes that give generous tax advantages to the investor (see Ch.3.3). In practice, although a number of schemes have been created, the whole programme is minuscule by comparison with the stock of housing and there is no real evidence that new investment in housing to rent can be created by these means.

The decline in the sector has now reached the point that it is difficult to find reasonable stock to rent in most cities, with the only property available being in run-down areas. Thus for many households in most cities and smaller towns the only available option is owner-occupation.

Council housing and housing associations Local authority housing (commonly known as council houses) was originally conceived as a suitable tenure form for the large majority of the population. The increased access to owner-occupation, the readily identifiable forms of property in this sector, which in several cities led to stigmatization of council estate residents, and the undesirable nature of much of the property created in the 1960s and early 1970s robbed the sector of any chance of fulfilling this rôle. During the 1980s the local authority sector was said to be in decline, with a reduction in the stock, a fall in the levels of repair and maintenance and the growing perception that local authority housing was accessible only to those in extremely poor financial circumstances.

Table 1.14 Sales in Great Britain under the "right to buy" legislation, 1980–92.

Year	Number of sales
1980	1,342
1981	84,897
1982	202,558
1983	144,456
1984	104,847
1985	96,486
1986	93,162
1987	108,033
1988	167,609
1989	185,791
1990	96,729
1991	53,557
1992	42,631

Source: Housing and Construction Statistics, 1979–92.

Government policy in the 1980s promoted the sale of council housing on favourable terms to sitting tenants, resulting in the change of tenure of much of the better-quality stock (see Table 1.14). Local authorities have not been permitted to finance significant quantities of new buildings, and expenditure on repairs and maintenance has been reduced to very low levels. Long waiting lists exist for council housing, though poorer properties are difficult to let. Access is restricted to those deemed to be in greatest need, which in practice implies families (especially one-parent) with young children.

In the local authority sector, a process of residualization is said to exist (see Ch.1.3). The existence of whole housing estates of the poor and dispossessed has become common, especially with the high level of unemployment that was the norm in the 1980s.

Housing associations have been considered by the present government as a suitable alternative to "politically inspired" local authorities. They are non-profit-making bodies whose sole concern is with housing. They obtain grant support from central government with which to finance the construction of houses, which they commonly let to households at "affordable rents". They have a good reputation as housing managers and have an active policy of involving tenants in decisions that affect their housing environment.

Level and trends of prices

Residential property Table 1.15 indicates the basic changes in prices in the housing market since 1964. The house price to earnings ratio has remained quite stable except for the periods 1972–3 and 1988–9, which were years of exceptional house price increases. The inflation in wages and prices has however led to generally rising house prices over the period, and buyers' real debts have fallen rapidly while their equity in the property has risen. This partly explains why owner-occupied housing has remained so popular in Britain.

House prices show wide variation between regions, as Table 1.16 indicates, though these have moderated since 1991. In the first quarter of 1991, the prices of houses in London, the Outer Metropolitan Area (OMA) and the South East began to fall, while the others were static or slowly rising. This process, which continued until mid-1993, corrected the over-rapid growth of house prices in the southern region during the house price boom of the late 1980s. The 1991 prices are important as they form the basis of the Council Tax valuations. In mid-1993, house prices were stable and the volume of sales had begun to increase.

Commercial property Table 1.17 illustrates two points. First, there is considerable difference between rent levels in different parts of the country. Such differences reflect demand and supply levels in each location. The

Table 1.15 Indices of housing data, 1964–92, (indices of actual prices).

Year	House price index	House land price index	Average earnings index	House price /earnings ratio	Retail price index
1964	100	100	100	3.26	100
1965	110	100	107	3.36	105
1966	119	117	114	3.4	109
1967	126	117	118	3.47	112
1968	137	133	127	3.51	117
1969	143	167	137	3.39	123
1970	153	183	153	3.25	131
1971	181	200	168	3.5	143
1972	248	350	189	4.29	154
1973	328	533	216	4.95	168
1974	333	533	256	4.25	194
1975	357	367	319	3.65	242
1976	383	367	368	3.4	282
1977	410	383	401	3.34	326
1978	481	483	457	3.43	353
1979	621	667	529	3.82	401
1980	717	883	647	3.61	473
1981	731	917	724	3.3	529
1982	754	1,033	785	3.13	574
1983	844	1,166	836	3.29	601
1984	951	1,316	908	3.42	631
1985	1,064	1,666	957	3.55	669
1986	1,143	2,133	1,041	3.52	692
1987	1,310	2,866	1,108	3.78	720
1988	1,446	4,200	1,209	3.84	756
1989	1,909	5,333	1,335	4.56	815
1990	1,910	4,500	1,461	4.18	904
1991	1,888	4,400	1,578	3.73	957
1992	1,823	na	1,664	3.50	993

Sources: Housing and Construction Statistics, various years, *Quarterly Digest of Statistics*, various years.
Note: na = not available.

quoted figures reflect rents for good-quality property; in each location there will be a range of poorer-quality property that will be competitively priced. Secondly, office rents are higher than those of industrial premises. Although there are obviously differences in building costs in the different sectors, it

Table 1.16 Regional house prices (new property), first quarter 1991, (£m).

region	Detached house	Semi-det house	Terraced house	All
UK	97,001	57,506	53,167	61,462
Scotland	83,214	55,498	na	47,986
Wales	102,037	59,911	50,240	56,637
N. Ireland	61,778	33,659	na	31,075
N	92,254	48,931	na	44,892
NW	94,990	51,219	47,751	62,762
Y & H	87,668	50,674	39,664	46,917
WM	94,893	48,618	45,067	57,702
EM	82,801	49,395	41,395	57,291
EA	87,421	47,116	41,805	64,922
SW	98,995	61,008	45,254	73,438
GL	140,296	83,269	80,603	68,887
OMA	127,527	83,602	77,168	63,528
SE	111,217	74,026	56,202	70,211

Source: Nationwide Building Society.
Notes: na = not available.
Key to regions: UK = United Kingdom, N = North, NW = North West, Y&H = Yorkshire and Humberside, WM = West Midlands, EM = East Midlands, EA East Anglia, SW = South West, GL = Greater London, OMA = Outer Metropolitan Area, SE = Rest of South East.

Table 1.17 Rent statistics, 1991 (£/m² p.a.).

Town	Region	Industrial premises	Office premises
London	Centre	na	305
Hounslow	London	118	288
Basingstoke	S.East	78	218
Bristol	S.West	70	205
Swindon	S.West	67	161
Ipswich	E.Anglia	54	169
Manchester	N.West	47	237
Newcastle	North	43	140
Sheffield	Yorks/H'side	43	143
Cardiff	Wales	50	156
Glasgow	Scotland	46	215

Source: Jones Lang Wootton (1991 figures).
Note: na = not available.

is not difficult to see how office developers are able to outbid industrial premises for land.

Table 1.18 Regional office and retail rent indices, 1980–91.

Year	Retail			Office		
	all	SE	N	all	SE	N
1980	70	76	53	74	72	86
1981	77	82	62	81	81	92
1982	81	86	68	85	84	95
1983	86	89	78	90	90	97
1984	96	96	96	97	98	100
1985	105	105	104	103	104	100
1986	117	120	115	113	112	103
1987	148	156	140	134	134	109
1988	181	191	165	185	181	133
1989	202	210	180	232	226	138
1990	202	202	187	261	252	218
1991	196	191	185	257	235	218

Source: Healey and Baker.
Notes: Data for December of each year. May 1985 = 100

Table 1.18 illustrates the general upward movement of office and retail rents with inflation. The boom period of the late 1980s is clearly identifiable, with retail rents in the South East rising by 81% from January 1985 (100) to May 1988 (181). Rents moved upwards most rapidly in the South East during this period, widening the gap between the South East and the rest of the country.

Currently there is evidence that the depression in both the office and retail sectors is much more severe in the South East than elsewhere and consequently the rent differences have narrowed (see Ch.9.1).

Another aspect of the boom in property prices in the late 1980s was the flow of money into property. As Table 1.19 shows, the total flow of money into property rose from a net figure of £2000M in 1980 to £14,200M in 1989. From 1988 to 1990 overseas investors were particularly important, with investors from Japan and Sweden dominating the figures. With the current depression in the market they are likely to be regretting these investments.

The figures for the bank debt of property companies (Table 1.19) show very clearly the extent to which the property boom was financed by borrowing from the banking system. Some of the money has been used to finance new developments that are currently unlet and some to fund the purchase of

Table 1.19 Flow of money into property, 1980–91, (£ million).

Year	Total money flow into property	Overseas invest-ment Into UK property	Bank debt of property companies
1980	2,000	100	4,800
1981	2,400	80	4,800
1982	2,800	120	5,000
1983	2,400	90	6,000
1984	2,700	80	7,500
1985	3,150	95	8,200
1986	4,200	150	10,300
1987	7,600	295	13,000
1988	10,300	1,900	19,000
1989	14,200	3,100	27,500
1990	8,600	2,900	37,000
1991			40,000

Source: Financial Statistics.

existing property. In either case the property companies concerned now have a high level of corporate gearing and the banks are concerned that a substantial proportion of their property debt may never be repaid.

Table 1.20 shows how property yields have changed in the South East and the North since 1982. The yield of 4.1 in the South East in 1982 implies that investors were on average, for this type of property, converting £1M of annual

Table 1.20 Property yields, 1982–91.

Year	retail		office		industrial	
	SE	N	SE	N	SE	N
1982	4.1	4.9	5.4	6.2	6.9	7.2
1983	4.3	4.9	6.0	7.8	7.5	8.9
1984	4.1	4.5	6.0	9.0	7.9	10.7
1985	4.1	4.5	6.3	10.0	8.8	12.2
1986	4.4	5.0	7.0	11.0	8.9	12.5
1987	4.7	5.1	7.5	11.3	8.8	12.5
1988	4.7	5.2	6.9	11.0	8.3	10.7
1989	5.1	5.7	6.5	9.9	7.8	10.2
1990	6.4	6.9	7.8	9.9	8.9	11.0
1991	6.7	7.4	9.4	11.6	9.3	11.6

Source: Healey and Baker & Hillier Parker
Key: SE = south east, N = north.

rent into a capital value of £24.39M (£1M multiplied by 100 over 4.1). This means they were typically paying £24.39M for each £1M of rental income. Equally it can be seen that yields in the North were higher and that a £1M rental income from retailing in the North would have been valued at £20.40M (£1M multiplied by 100 over 4.9).

In 1982 there were low rates of interest in the economy and Britain had started to move out of the depression of the early 1980s. Retailing and office markets boomed and yields remained low. This meant that the rapidly rising rents in each sector identified in Table 1.18 were translated directly into increases in capital values. By 1990 the economy was moving into depression and interest rates were rising, so yields were generally rising. As rents had stopped rising, this meant that the investors were actually prepared to pay less for a given rental income than previously, i.e. capital values were falling. The process continued into 1992, in that yields were rising and property values falling. For properties let on long leases the contract normally does not permit rents to fall, but for unlet office property in London and the South East the rents they are likely to obtain are also falling. This implies even faster falls in capital values.

Table 1.21 Commercial property construction costs, 1991 (£/m²).

costs	Building
factory	237–322
factory/office high technology	
shell & core	429–565
warehouses	237–288
offices	
basic:	
non air conditioned	622–904
air conditioned	893–1,706
prestigious:	
high rise	1,806–2,486
shops (shells)	
small	429–542
large	396–452

Source: Spons' Building Cost/Price Book, 1991.

Since October 1992 there has been renewed interest in property investment, because with low property prices and low borrowing costs an
investment in property looks attractive. Properties that are not let are,
however, very difficult to sell.

Construction cost data are normally obtained from quantity surveyors, who
will provide estimates for each "unique" building. The figures in Table 1.21
indicate typical levels as of spring 1991.

Service charges Where a single property is occupied by a number of businesses, or the landlord provides services to the occupier for any other reason,
the cost of those services is recouped by the landlord by charging the tenants
a service charge. The size of the service charge is clearly directly related to the
costs incurred by the landlord and is not intended to be a separate profit centre.

The overall consequence of the service charge and the requirement of
tenants to pay for all insurance and repairs is that the rent income to the
landlord is a pure flow of income, a return on the investment.

Requirements for land

Very little information is collected in Britain on land requirements, though a
number of surveys provide evidence of past trends. The housing market has

Table 1.22 Land changing to residential use
by previous use, 1985–8.

region	Previous use		Area (ha)
	rural (%)	urban (%)	
North	58	42	1,490
Yorks/H'side	52	48	2,490
E.Midlands	67	33	3,320
E.Anglia	63	37	2,180
S.East	46	54	9,570
S.West	68	32	4,750
W.Midlands	55	45	3,240
N.West	41	59	3,060
England	55	45	30,100
Wales	67	33	1,870
Scotland	59	41	3,610
Great Britain	56	44	35,580

Source: Housing and Construction Statistics; DOE.

Table 1.23 International comparisons of dwelling space, 1976.

Country	Dwellings per '000 inhabitants	Average usable floorspace per dwelling (m²)
Austria	388	86
Belgium	386	97
Denmark	424	122
Finland	318	71
France	399	82
Ireland	259	88
Italy	329	na
Netherlands	348	71
Norway	365	89
Portugal	na	104
Spain	344	82
Sweden	394	109
Switzerland	424	98
United Kingdom	388	70
West Germany	407	95
Canada	323	89
United States	333	120

Source: United Nations 1976
Note: na = not available.

Table 1.24 Floorspace per worker in commercial premises in London (m²).

	average	range
office	20.26	10.24– 29.63
retail	43.89	30.09– 54.20
industry	38.02	27.00– 48.36
warehousing	61.41	31.59– 90.90
retail warehousing	94.33	na

Source: R. Barker, London Research Centre 1992.
Note: na = not available.

been most extensively studied, and, as Table 1.22 indicates, there was a considerable loss of rural land to housing in the mid-1980s. The significance of the South East region in these figures illustrates the particular boom in property prices and development activity there. Other data show that very little land has come from green belts, which implies that much of the land must have come from areas away from urban conurbations where green belts do not exist.

The Department of the Environment (DOE) estimates the urban area per person to be $288\,m^2$ (1981), and predicts this to rise to $310\,m^2$ by 2001. The total urban area is expected to rise from 10.2% in England to 11% by 2001. The later figure is consistent with estimates from other countries (see other volumes in this series), but precise comparisons cannot be made because of definition problems.

The DOE estimates that the density of new housing developments is 30–40 per hectare, a figure that is consistent with house-builders' comments.

Very little evidence is available on living space per person in Britain. The United Nations figures given in Table 1.23 are the most recent ones readily available. Since 1976 there has been an expansion in the number of houses in Britain: now the figure is 403 dwellings per 1000 persons. This includes a considerable number of houses that have been declared unfit for human habitation.

Since 1975 there has been some investment in smaller homes, which are seen as more suitable for the smaller households that are now more common in Britain.

Figures for space per person at work are not collected centrally. Some survey evidence is available and one example is reproduced in Table 1.24. Although the figures refer to London, they are unlikely to be very different elsewhere.

Each modern building is designed to specific standards, which include an expectation of the number of workers, but these standards vary considerably. There is a widespread recognition that a workplace revolution is taking place in Britain. The main characteristics of this are:
- shorter working hours
- more shift work
- job sharing
- flexible working hours
- more second jobs
- a growing black economy
- factories dominated by robots
- six weeks' paid holidays
- better working conditions
- worker education and retraining

– employment located away from the cities.

Each element in the change in employment practice has implications for property. Developers tend to respond to current perceived demands rather than to analyze all of these changes systematically.

1.5 Trends in spatial development

National trends

The two key issues to recognize are the north/south divide and the urban–rural shift. The latter, especially in the context of land and property markets, concerns changes in the structure of agglomerations, peripheral urban development and the strong growth of small and medium-sized self-standing towns.

Since the 1930s, at least, the dominant movement of population and employment has been from north to south. This represents a reversal of the prevailing trends during the 19th century and a return to the broad distribution found before 1750. This north to south movement declined substantially in the late 1960s and early 1970s, possibly reflecting some regional policy effects at that time, but became very firmly re-established during the 1980s. These extremes are illustrated in Table 1.25.

Interregional changes have certainly been a factor in the minds of many of the actors in the land and property markets in the UK, but this broad level of generalization disguises many changes visible at the subregional level of analysis (see below). Prosperity and population growth may be found in many parts of the north, including the Grampian Region of Scotland (see Ch.6.3); while the opposite is to be found in several London boroughs and in 1992–3 more widely in the South East of England.

Different definitions of the north and south, and of the location of the boundary between them, are to be found. Table 1.25 takes the south to include the four most prosperous standard regions of the UK: East Anglia, South West, East Midlands and South East. For many purposes, London is treated separately from the Outer South East, and it is within the latter that the greatest pressures of population and development have occurred in recent years. The far South West (especially Cornwall) does not share the regional level of prosperity, having high levels of unemployment and a highly seasonal economy based on tourism. The north is taken as the remainder of the country, not strictly northern in the geographical sense. Indeed discussion of the issue has often centred on the question of where to draw the line between the north and the south on the basis that growth and prosperity are to be found in the south, and a lack of these in the north, so the boundaries are redrawn so as to include all areas of growth in the south. The concept of a

42

Table 1.25 Population change in Great Britain by standard region, 1984–87, (annual rate per '000 people).

Region	1974–77	1984–87
South		
East Anglia	9.9	12.7
South West	5.5	9.5
East Midlands	3.7	5.9
South East	(2.4)	4.0
South Total	0.5	5.8
North		
Wales	1.8	3.4
West Midlands	(0.8)	1.4
Yorks/H'side	(0.4)	(0.3)
North West	(3.2)	(1.3)
Northern	(0.5)	(1.8)
Scotland	(0.9)	(2.2)
North Total	(1.1)	(0.4)
GB	(0.3)	(2.7)
Difference N–S	(1.6)	(6.2)

Source: Champion and Townsend (1990), calculated from mid-year estimates.
Note: Regions ranked by 1984–7 change rate.

divide, rather than a slope from south to north, is thus embodied in the thinking of those concerned with property and economic development – with discussion from time to time about how far north the dividing line has moved.

It can be seen that the net movement north to south rose from a low of 10,000–20,000 per annum in the early 1970s to a peak of nearly 70,000 per annum in the mid-1980s. This figure has since declined (see Table 1.26). These flows do not consist of a cross-section of the population; they have included disproportionate numbers of the younger, more qualified and dynamic sections of the community. According to the 1983 *Labour Force Survey*, the bulk of the movement to the South East consisted of non-manual workers, although there have been shortages of manual workers in the South East and higher unemployment among manual than among non-manual workers in the North. One of the reasons for this has been the property market: lower earners have found it extremely difficult to obtain suitable accommodation for their household in the south even though they may have been successful in obtaining a job.

Table 1.26 Migration within the UK in 1991, (thousands).

region of destination	region of origin											
	UK	N	Y&H	EM	EA	SE	SW	WM	NW	W	S	NI
UK		49	85	81	48	265	99	88	100	47	47	9
N	50		9	4	2	14	3	3	8	1	5	
Y&H	85	9		13	4	24	6	7	15	3	4	1
EM	90	4	15		6	31	6	12	9	3	3	
EA	58	2	4	7		32	4	3	3	1	2	
SE	223	14	22	25	23		49	26	28	15	17	3
SW	121	3	6	7	4	65		13	9	8	4	
WM	83	3	7	10	3	28	11		11	7	3	1
NW	90	7	14	8	3	27	7	11		7	6	1
Wales	51	1	2	3	1	17	8	8	9		1	
Scotland	56	6	5	4	2	20	4	3	6	2		2
N. Ireland	12		1	1		6	1	1	2		2	2

Source: Patient movements recorded by the National Health Service.
Key to regions: UK = United Kingdom, N = North, Y&H = Yorkshire and Humberside, EM = East Midlands, EA = East Anglia, SE = South East, SW = South West, WM = West Midlands, NW = North West.

Regional policy

There continues to be a national regional policy, in the form of incentives for development in particular sectors, especially manufacturing, in designated assisted areas of the country.

The map of assisted areas was redrawn at intervals during the 1980s, reducing the total area and population covered and targeting the benefits more precisely. In recent years, this policy has been of particular significance in the context of EC regional policy and the designation of areas of benefit for the three EC structural funds – the regional development fund (ERDF), the social fund (ESF) and the guidance section of the agriculture fund (EAGGF).

Since the three funds were co-ordinated around a common set of objectives in 1988, only Northern Ireland has qualified as an Objective 1 region, the assisted regions of mainland UK being Objective 2 regions. However, Merseyside and parts of the north of Scotland may also become Objective 1 regions as part of the reform of the structural funds proposed in 1993.

In many of the major cities in these regions, financial benefit under inner-city and urban development policies also applies and may in many cases be of greater significance as far as the property development process is concerned (see, for example, Ch.6.2 and Ch.9.3).

Property values

House price data are very often seen as an indicator of the north/south divide (see Ch.1.4). The phenomenon of a widening of the gap between north and south during the 1980s was very strongly reflected in both residential and commercial property markets. Increased differences in property values are attributed to a number of factors: differences in interregional economic performance both overall and between different sectors of the economy, the south benefiting from the greater presence there of growth sectors such as financial services (see Ch.9.1), high-technology defence industries, computer industries and commercial services (see Ch.6.1); population movements and demographic changes; and the application of the free market ideology of the government to planning policies.

All forms of regional economic planning were abandoned during the 1980s, and spatial planning at the strategic scale has been greatly weakened (Healey et al. 1988). At times the government was moving towards substantial relaxation of restrictive policies such as the green belt and reduction of the rôle of county structure plans. At the same time, development was encouraged and actively promoted, especially in the Docklands (Brownhill 1990). A property boom, coinciding with the height of the disparity of property values between north and south around 1987–8, possibly represented the peak of the north/south divide in the 1980s.

Subsequently, prices have dropped in the south, leading to some convergence nationally, and the government is now taking a more positive attitude to the question of regional and strategic spatial planning guidelines and firmer local planning policies. These issues are discussed in Chapter 3.1 and illustrated by Chapters 6.3 and 9.2.

European spatial context

In 1994, the UK will be linked to continental Europe by the fixed rail link through the Channel Tunnel. This is likely to have profound implications for the development of spatial policy in the UK in ways that are not yet widely understood or appreciated. Associated investment in the inter-city rail network has not yet taken place in the way seen in France. Although the rail links north of London are relatively good, and, in the case of the London—Leeds—Newcastle—Edinburgh main lines, have recently been upgraded to allow speeds of up to 250 km per hour, no final decision has yet been taken on the form or route of the crucial missing link in the high-speed rail network between the Tunnel and London, and on the connection with the rail network north of London. Work cannot now be completed until at least 1997. The consequence will be a reduction in the spatial distribution of the benefit of the Tunnel throughout Britain, and pressure for development, for example for distribution facilities, in the southeast corner accessible to the Tunnel.

Agglomeration trends

During the period from 1945 to the mid-1960s the large cities were seen as the growing foci of commercial development and the "natural" location of both business and housing. The process of decentralization away from the urban cores had begun before this date, but about this time it became apparent that both housing and commercial activity were relocating to new environments.

Public policy, particularly planning policy, had encouraged the trend with the creation of out-of-town housing estates and the new towns policy. The green belt policy, which surrounded cities with a belt of land where development was not permitted, limited the extent of peripheral development, especially in London. Small towns in accessible locations began to grow rapidly as both individuals and companies saw the purpose-built environment in these locations as more suitable for their requirements.

The main agglomeration trends can therefore be identified as: suburbanization, commuting, the growth of small towns, and inner-city decay (Champion and Townsend 1990).

Suburbanization In general, urban areas have grown in size with the creation of suburban zones on their periphery. Such zones consist primarily of housing but with the addition of local services, particularly retailing. Much of the housing in the 1980s was for owner-occupation, but during the 1960s and 1970s local authorities often located council housing estates on the urban periphery.

The occupants of these housing developments depend on a limited range of local services but commonly require cars for access to a wider range of facilities. In addition, the main source of employment for these people has been the urban core.

Commuting A corollary of suburbanization has been the need to travel some distance to work. The commuting practice for much of the post-war period has been a radial pattern to and from the urban core, but in the 1980s there was evidence that commuting was becoming more complex, with people increasingly making journeys across segments of the urban area as the location of employment changed. Commuting is restricted by time and cost. Most urban economists and property developers recognize an urban travel-to-work district that extends away from the urban area, particularly along good transport routes.

Where the costs of housing are very high, as in central London, it is possible to find people commuting from some considerable distance. The evidence of the housing market is that the maximum commuting distance house purchasers are prepared to contemplate is determined by a 100 minute travel-time band or a maximum commuting time of 3 hours and 20 minutes per day. Given that some trains on the inter-city rail lines carrying these commuters travel at up to 225 km per hour, the effective commuter belt can reach a long way from the city centre. At the height of the boom of the late 1980s and the north/south divide in property prices, the 100 minute threshold was, for example, reflected in housing developments as far north as Doncaster on the London–Edinburgh line for the London employment market.

On the other hand, the southeastern suburban rail network is carrying loads well beyond its design capacity, and is widely acknowledged to be drastically in need of investment and improvement.

In most urban areas the dominant mode of transport is the car, and the post-1945 road-building boom has encouraged people to live further away from their place of work. There is little doubt that road congestion is rising in Britain and consequently that car commuting is becoming less congenial. It is nevertheless a widely adopted practice. This is especially true in the South East because of the inadequacy of the suburban rail system (London Transport and British Rail Network Southeast) and because of the spatially diffuse distribution of employment. The use of the car and the growing congestion of our cities has encouraged property development to take place

47

at non-central locations. The M25 radial motorway around London is both a cause of and a response to this situation.

The growth of small towns   Since the mid-1960s and most particularly during the 1980s, the fastest-growing localities in terms of both housing and employment have been small towns with a population of less than 100,000. The decline in the traditional manufacturing process, where a very large number of workers was needed for assembly-line production, and its replacement by smaller industrial units and more flexible work patterns have enabled commerce to move away from its traditional locations. The availability of telecommunications facilities and access to the nation's motorway network have solved problems of access to suppliers and customers. In addition, development on greenfield sites on the periphery of these towns is far easier and cheaper than on restricted inner urban sites.

For employees, the environment obtainable within these towns, especially the access to the countryside, has been seen as far superior to that obtainable in what are perceived as decaying urban areas. The towns are perceived as islands of prosperity, of good schools and health facilities and of traditional middle-class values. They are therefore highly attractive to those who value this lifestyle.

Inner-city decay   Within most British cities the problem of inner-city decay had been observed by the late 1960s and had become the subject of active public policy by the early 1970s.

The problem has been identified in two ways. First, inner-city decay has been associated with concentrations of people who are poor and in need of social provision. Concentrations of unemployment, poor health and educational achievement, one-parent families, homelessness and racial minorities in inner-city locations have developed. A process of residualization has occurred whereby the financially better-off have left these areas to the poor, who are unable to move or improve their lifestyle. Secondly, inner-city decay is associated with a deterioration in the physical fabric of the city. Poor-quality council and private housing is in an increasingly worsening state of repair, school and hospital buildings need improvement and much derelict land exists.

The two issues are not unconnected because the private sector has been unwilling to invest in locations where its developments and property values would be blighted by the existence of nearby dereliction, and the absence of private investment has led to a steady decline in the area. The physical decline and the associated decline in employment opportunities have led to further social decay. In addition, the public sector was not a heavy investor in the urban areas in the 1980s, and in particular there have been large cuts

48

in the public sector house-building programme as the government has increasingly relied on the private sector to meet housing need.

It would be wrong to assume that all urban areas are equally subject to decay or that the problem afflicts the whole of an urban area. Parts of most cities are prosperous and physical regeneration continues. There are, however, substantial and growing areas where these problems can be clearly identified and have been addressed by successive governments in a wide variety of ways, through inner-city and urban policy initiatives.

Inner-city policy is outlined below, and policy instruments are outlined in Chapter 3 and illustrated by the case studies in Chapter 6.2 and Chapter 9.3.

Inner urban development

In Britain, two terms are commonly used to describe central urban areas: city centre, or CBD; and inner city.

The city centre or central business district (CBD) consists of the central retail core, which historically evolved to meet the major retailing needs of the urban region, and the prestigious office centre. In all urban areas except London, this core can be easily identified and spatially defined. Today it is usually commercially viable and often the subject of commercial regeneration. In all major conurbations there are now significant secondary centres of retailing and office space that are themselves commercially viable.

In London, the scale is such that a number of significant centres have developed with the growth of the city. Prime retailing sites can be found outside the city centre and some office decentralization has occurred.

The term "inner city" commonly refers to the run-down urban areas that can be found in all of Britain's major conurbations around the core. These are predominantly areas developed in the 19th century, with housing associated with the industrialization of that time, now displaying economic and social problems described above. Similar social conditions are found on many post-war council housing estates, many of which are located on the periphery of cities. Such places are also nevertheless often the focus of inner-city policies and therefore the definition of the inner city is understood to include such outer housing estates.

The city centre The city centre is conventionally recognized in policy as a commercially viable location and as such it is not traditionally seen as being in need of direct financial intervention by government, although there are exceptions. During the 1980s three types of development should be noted.

Retailing Retailing in Britain was generally prosperous throughout the 1980s, though in recent years there has been a considerable decline in its viability, especially in the south. The pattern of retailing has been undergoing major changes as new developments on the periphery of the city have been created (see below). Some questions have been posed about the viability of the city centre (Distributive Trades EDC 1988; HMSO 1992), and these have been translated into policy guidance to local authorities to pursue planning policies that are consistent with the maintenance of a viable city centre.

In general, city centres are surviving as these planning policies are supported by the retail industry, so many of whose property assets are dependent on sustaining the value and viability of the core. In practice, the evidence suggests that the central retailing areas have not so far been severely damaged by peripheral developments. They appear to retain their attractiveness for comparison shopping, that is, for shoppers wanting a wide variety of products and services, and for the purchase of higher-order products (electrical products, expensive clothing, furnishings, etc.). There is some evidence that within the city centre the most prestigious locations are very prosperous and physical regeneration by the private sector is continuing, but that more peripheral or marginal locations are showing signs of decline.

One particular change in the 1980s was the decline in the use of partnership arrangements between local authorities and commercial developers. This reflects the more localized areas of redevelopment in the 1980s (except for Enterprise Zones and Urban Development Corporation areas) in comparison with the large-scale city centre redevelopment of the 1970s.

Outside of the city centre, the local district shopping centres show clear signs of decline, particularly in the inner-city locations. It appears that shoppers use either the city centres or the peripheral developments for their requirements. Many district centres contain empty property and some is falling into severe decay.

Office space Throughout the 1980s there was a growing demand by property users for office space. This trend was particularly strong in central London (see Ch.9.1), but it could be found in all major urban areas. Office space development has been seen as the responsibility of the private sector and therefore outside the direct influence of government. Offices tend to be located in the city centre and workers commute from good-quality suburban housing.

During the 1980s some decentralization occurred from London to other city centres and from city centres to peripheral locations. It is argued that these decisions reflected the cheapness of peripheral land and hence development costs and rents and also the wish by many workers to avoid the need to commute to the city centre on expensive and congested transport routes. They also reflect the economic pressure to use the most expensive space for high-value office functions, decentralizing more labour-intensive routine office functions to lower-rent locations. Changing technology, particularly information technology, has enabled firms to locate secondary office functions away from city centre head offices with minimal loss of efficiency.

In the economically poorer regions, office development is highly cyclical, with commercial regeneration occurring only at the peak of the country's economic boom. The case study of Newcastle's Business Park (Ch.6.2) provides such an example. It also points to one of the changes in the 1980s, in that the government's spatial regeneration policies, Enterprise Zones and Urban Development Corporations, influenced the location of office development in some cities by means of subsidies.

The boom in the construction of office space generated at the end of the 1980s has now led to excessive supply in relation to demand, especially in London where the most prominent case, that of Canary Wharf, stands as a symbol of the excesses of the boom (Barnes 1990). It is inevitable that it will be several years before the situation finds equilibrium again and new building is viable. In the mean time, there is little prospect of physical regeneration through office development by the private sector.

The service sector One of the major changes in Britain during the 1980s was the switch from manufacturing employment to service sector employment (see Ch.1.2). The location of much of this activity is at the city centre and indeed has been the cause of much of the growth in demand for office space. Other forms of space are also required, and much traditional industrial property has been refurbished to provide suitable premises.

The service sector creates new demands on infrastructure, telecommunications, road and rail transport, etc., and some consideration is now being given to these within the planning system.

The inner city The inner-city problem was initially identified during the later 1960s as a decline in the physical and social conditions within a segment of the city (Lawless 1979). Some policy initiatives aimed at localized physical repair and social support attempted to remedy the situation. These policies failed to recognize the extent of the relocation of people and jobs away from the urban core and consequently were only partially successful. In the 1970s, a new analysis of the problem, related to the structure of the national economy, became more widely accepted, and a specific link was made with the new town programme. It was argued that the latter had diverted both public financial resources and the more enterprising and educated human resources disproportionately from the inner cities to the new towns. Following this line of reasoning, the end of the new town programme was announced in 1976, the rundown being largely complete by 1988.

During the 1980s, the extent and depth of decay deepened, and even in the capital city there are widespread areas of social and physical decay, within which the property sector does not function well. The poor image of these areas reduces local property values, making property development non-viable, and consequently they are worsening over time.

In the housing sector, few properties are owned by their occupiers and those that are owned are often in a poor state of repair (Karn et al. 1985). The owner may find that, even if they can afford to repair the property by borrowing the necessary finance, the increase in value of the property after the repair would be less than the cost of the repair and consequently the banks are unwilling to provide a loan. Those who can afford to tend to move elsewhere, so the occupiers tend to be those who are unable to move elsewhere.

Property values also mean that it is generally impossible to build new houses at a profit. A comprehensive redevelopment of a large area, creating a whole new environment, may be viable, but this is normally outside the scope of most developers' activities without public financial underpinning. Private sector developers would find land assembly of a sufficiently large site very difficult, and neither tenants nor investors are prepared to consider such locations.

Commercial businesses in these areas do exist, but they tend to provide low-paid and part-time employment, occupy old and poor-quality premises and have little incentive to invest in new property. Buildings are generally of poor quality and in need of repair and improvement. These areas do not attract the main institutional investors, as this is not prime property, and tenants and occupiers cannot be relied upon to fulfil the terms of conventional leases.

Generally the point must be made that such areas have declined beyond the point where private sector regeneration would be commercially viable. The areas are subject to cumulative decline, which could be arrested only with public sector intervention.

During the 1980s, while expenditure on amelioration of social conditions grew rapidly, policy has been primarily directed at improving the physical fabric of the area with the use of public subsidies. The central thrust of government policy was to encourage the physical redevelopment of the cities by the Enterprise Zone (EZ) and Urban Development Corporation (UDC) policies (see Ch.3.3). These have attempted to solve the problem of the commercial non-viability of development by providing subsidies for either the reclamation of derelict land, the provision of transport infrastructure, property development and/or property occupation. The case study of Newcastle's Business Park (Ch.6.2) illustrates the general approach. The social problems have been addressed but have not received the priority devoted to physical regeneration, as the implicit theory has been that social benefits can be expected to "trickle down" from the physical regeneration of the city to the local community.

The UDCs have been particularly active in land assembly, derelict land reclamation, improving infrastructure and subsidizing the development process (see Ch.6.2). They have attempted to improve radically whole areas and it was anticipated that the benefits of such regeneration would trickle down to the social conditions of the local inhabitants of the area. In practice they appear to have had little, or negative, social impact, although they have resulted in the creation of new built environments and new employment opportunities.

One particular difficulty associated with the physical regeneration of urban areas has been that it has created employment opportunities for highly skilled workers yet the local labour force is dominated by unskilled people. This issue has been most acutely observed in London, but more generally it does point to one of the weaknesses of trying to "solve" urban problems by physical regeneration alone.

Peripheral urban development

The periphery of urban areas has been the subject of extensive development pressures since 1945. Every city has expanded spatially towards the limits created by the green belts or other restrictive planning policies established since the 1950s (Hall et al. 1977, Elson 1986). All forms of development have occurred, though the predominant land user has been housing. Where development pressures have been particularly strong, building has taken place beyond the green belt. New towns were themselves situated in such locations, and are, for example, important features of the settlement pattern of Hertfordshire, referred to in the case study in Chapter 9.2. Such settlement has subsequently become incorporated within London's commuting pattern.

Housing One of the most obvious characteristics of urban development in Britain has been the process of suburbanization. Low- to medium-density housing has been created for owner-occupiers by the building of detached and semi-detached houses, each with their own garden. Such housing has found a ready demand during most parts of the property cycle. Within the housing estates, the developers typically provide a parade of 5 to 10 shops to meet local needs, but otherwise the area is devoid of other forms of development (see Ch.6.3).

Occupiers are typically earning average or above-average incomes and own a car, which is commonly used to commute to work. Local authorities provide schools and other facilities over time as the area becomes established as a zone of housing.

The extent to which the suburban zone extends away from the urban centre depends upon the transport infrastructure. Road transport is important as, with the decentralization of employment away from the city centre, the pattern of commuting has become more complex, especially where high-capacity radial routes exist. The most notorious example of this is the M25 motorway around London. Each of the residential case studies (Chs 6.3, 9.2, 9.3) illustrates forms of peripheral development.

Retailing It is conventional in Britain to talk about waves of innovation in retailing, each with its own spatial implications. The first wave in the late 1960s and 1970s saw the creation of large single-occupier multi-product stores built on the periphery of urban areas or adjacent to good road infrastructure and within easy reach of a large number of people. The store typically sold food and many other items such as furnishings, paint, wallpaper, electrical products, etc. The store was cheap to build because of the availability of large undeveloped sites, and car owners could be attracted by the provision of free parking adjacent to the building.

The stores typically opened into the evening and particularly catered for those who wished to shop by car once a week or even once a month. Public transport was generally not available to these locations and consequently they were not accessible to the poorer members of the community.

Such stores were not seen as particularly threatening to city centre retailing, for at most their impact was on food retailing, which had already relocated away from the centre and could, arguably, be more efficiently provided at these new locations. In particular, the problem of transporting bulky food items from store to car seemed to be solved in a way that was almost impossible to achieve in the city centre.

The second wave of retailing, retail parks, started in the early 1970s and was most popular in the 1980s. These were created on the periphery of urban areas and adjacent to good road transport systems as they sought to attract car-born shoppers. Goods with high volume to value ratios are particularly suited to this form of development.

The typical development includes 5 to 10 retail units and often does not include food retailing. Typical occupants are car accessories, flat-pack furniture, do-it-yourself (DIY), electricals, toys and more recently some services such as banks. The properties are little more than warehouse units with a retail frontage. Their low construction costs enable low rents per square metre to be charged, which is an important consideration for many of the users who need a large floorspace for their products.

Retail parks in the 1980s were strongly dominated by national retail chains, with the result that many of the developments contain the same tenants. With the growth of retail spending in the 1980s this did not matter because there was little evidence of saturation of the market, but in the more competitive conditions of the 1990s there is little doubt that some of the developments are far less attractive than others and that it may be necessary for the owners to consider modifications or even redevelopment.

The response of the planning system during the 1980s was to recognize that this form of retailing was highly efficient and attractive to shoppers, especially car-born shoppers who could take home the bulky products immediately. In addition, the government insisted that the planning system was not to be used to protect existing retail forms and consequently the form of development was often accepted. Two major problems were identified by planners. First, many of the best sites were located in areas of green belt or countryside and consequently development was resisted. Secondly, planners were concerned about the implications for the viability of city centres of the widespread development of retail parks. The issue has been hotly debated but few guidelines have currently been created by central government and therefore a consistent policy has been lacking.

The third wave of retailing has been the creation of peripheral shopping centres that aim to reproduce the diversity of shopping experience found in the city centre while also providing an enclosed environment and access by the motor car.

The MetroCentre in Gateshead, Tyne and Wear, opened in 1986, was the first and one of the most famous developments. This is one of the case studies discussed by Pfeiffer in the Annex. The concept has been closely followed by the Meadowhall Centre in Sheffield, which like the MetroCentre is located on a former derelict industrial site away from traditional shopping locations in an area benefiting from inner-city urban development programmes.

The MetroCentre remains the largest purpose-built shopping centre in Europe, with 360 shops, over 30 restaurants, a 10-screen cinema, a bowling alley and an indoor funfair. It provides over 12,000 free car parking spaces plus additional coach parking facilities and is also serviced by over 100 buses per hour and several trains a day. The site, about 5 km from the city centre, is adjacent to the A1 road, which is the major east coast north–south route, and it also forms part of the Tyneside Enterprise Zone, which enabled the developer and the occupiers to receive public subsidies. In the case of this development, the simplified planning regime of the EZ was also of benefit to the developer.

The property is now owned by the financial arm of the Church of England, the Church Commissioners, whose agents intensively manage the development so as to achieve a safe and secure shopping environment and to ensure that the development remains a commercial success. The non-retail functions, particularly the leisure facilities, are seen as highly compatible with the shopping facilities and capable of attracting people to the centre for their sole use. In this way the development is seen at least partly to mirror the city centre, where retailing is only one of the facilities.

The development attracts shoppers from throughout the region and from other regions. It is therefore seen by planners as providing a regional shopping rôle and as such it is in direct competition with the city centre. One view of the impact of the MetroCentre is that it has stimulated competition from the existing regional centre, Newcastle's city centre, reinforcing the quality of the latter and leading to an improved shopping environment for all shoppers. Another view is that, while the most prestigious shops in Newcastle will survive, the marginal shops will decline and the city centre will in effect shrink in size.

Industry Industrial relocation away from traditional central city sites began in the late 1960s but it was only in the 1980s that major pressures from industrial developers on the periphery of urban areas were felt (Fothergill et al. 1987).

Modern developers of industrial units have, especially since the change in the Use Classes Order (see Ch.3.1), sought to create low-density developments in attractive environments away from the urban core. Sites adjacent to London's orbital motorway, the M25, have been in particular demand as they are attractive to employees, who can live and work away from the city centre, and also offer the firm access to the national motorway network.

Proposed developments may be on land that is not designated for industrial development and consequently many potential developers have resorted to planning gain agreements to enable their proposals to achieve planning permission. Furthermore, many of the developments are only loosely related to industry and are really office parks (Wood & Williams 1992). The case study of Stockley Park (Ch.6.1) is an example of this form of development.

CHAPTER 2
The policy environment

The election of the Conservative government in Britain in 1979 marked a significant change in the policy environment in which land and property markets functioned. Prior to that date the system could have been said to be planning led, with major initiatives such as new towns and the regional dispersion of industry originating in the public sector. After the property price boom of the early 1970s there was a determined effort by the 1974 Labour government to control the development process by the effective catchment of all increases in land values for the public purse. The instrument for this purpose was the 1975 Community Land Act, which enabled local authorities to purchase all the land that was likely to be required for development for the next 10 years. This Act was never fully implemented because it required considerable public finance, and the later 1970s saw Britain with worsening economic fortunes. The result was that Britain's property sector in the late 1970s was being led by a government committed to social control but without the means to implement its policy.

The incoming Conservative government in 1979 was committed to a radically different philosophy. It was pro-market and pro-development and in a famous speech by the then Secretary of State for the Environment, Michael Heseltine, the public planning system was castigated for "locking up jobs in the filing cabinet". The government was also innovative in that it looked for radical new approaches to problems and was prepared to commit funds where it saw the outcome would be beneficial for the private sector.

The government has since 1979 not only encouraged market-led development but seen it as the driving force in the regeneration of urban areas. While social problems have not been ignored, there has been a tendency to promote property-led physical regeneration as the solution to urban decay (Healey et al. 1992). A number of key policies were developed that have resulted in public policy having a certain form and direction that is now familiar to the property sector.

2.1 Market-determined development

The prevailing view after 1979 was that the market knows best what form of development society requires. Commercial developers build the type of properties for which people are prepared to pay and the rôle of the public sector is primarily one of fácilitating the process, or of becoming enablers.

For example, it became no longer sufficient for local authorities to ensure that sufficient land was available for housing; it also had to be available in the form and location that purchasers require.

Additionally local authorities must ensure that the land to be developed is free from ownership or other constraints. In reality it is impossible to ensure that everyone can purchase the form of house they require in the location they desire, but the principle has been set that household demands should be the guiding force and that planning control should facilitate the process as much as possible.

In this policy environment, social needs that necessarily fall outside the parameters of ability and willingness to pay are discounted and, where the public sector is unavoidably involved, the decision has been made by reference rates of return. Such approaches attempt to mimic the market decision and ignore any wider social variables. This market-oriented thinking has also led to substantial interest in the concept known as planning gain or exaction and to growth in the use of developer agreements. Such agreements, now termed "planning obligations" in the Planning and Compensation Act 1991 (see Ch.3.1), in which developers and local planning authorities reach an agreement whereby the award of planning permission is accompanied by an obligation on the part of the developer to provide certain social or physical infrastructure, are now widely expected.

Another example of the policy change was the decision to abandon the restrictions on office development in London and to weaken controls on design, plot density, etc., with the result that London now has a large supply of land that could be used for office development if the market so dictates. The problem at present is more one of over-supply, however (see Ch.9.1).

The location of development

The pro-market stance of government policy resulted in location decisions being left primarily to market forces. Industry and commerce were allowed to locate where they wished, and local authorities were encouraged to compete to attract mobile firms. Some influence remained with government and the EC Commission because of the continuation of regional capital grants and the operation of the EC structural funds. The high cost of land in preferred locations further modified business decisions.

In a number of locations that had been particularly associated with the decline in the coal and steel industries additional finance was made available to provide sites and infrastructure and to subsidize incoming firms. However, the overall area and population covered by regional assistance designation of this sort was rapidly reduced in the early 1980s from about 40% of the population in 1979 to under 25%. Most recently, some extension of the designated area is in prospect with the reform of the EC structural funds being undertaken during 1993.

The major break with the pro-market stance, however, is in the context of the issue of urban regeneration. The government wished the private sector to be the driving force in re-creating jobs and housing in inner urban areas but it recognized that the private sector would not become involved because the developments were not likely to be profitable (see Ch.4.1). Initially the main policy was that of Enterprise Zones (see Ch.3.3) where the investment in buildings was subsidized. After the 1987 election, inner-city policy became more important, the Development Corporation experiment was extended and the City Grant system expanded. In the latter two cases the primary principle has been that of "gap funding" (see Ch.4.1).

Privatization

For the property sector the privatization programme has led to a number of new private sector firms with large land holdings. The largest are the new water companies, but much of their land is in rural areas. Other privatized companies such as the bus companies and British Leyland were able to exploit their land holdings successfully during the boom years in the property market at the end of the 1980s but are now unable to do this.

For those industries that have remained up to now in public ownership (particularly British Rail), there has been a strong pressure to exploit their land holdings in the same way as a commercial firm would. Again this has stopped in the current land market conditions.

Planning gain

Planning gain is explained in Chapter 4.3 below. It is important to recognize that, while government has not explicitly encouraged this form of bargaining between planning authorities and developers, it has increasingly condoned rather than discouraged it. Some local authorities have certainly taken advantage of this position to extract community gains from developments that they may well have rejected on purely planning grounds.

Recently the government has introduced new guidelines that appear to encourage developers to offer planning gain to the local community.

Marketing the city (image policies)

A number of cities have attempted to promote themselves to tourists, industry and their own inhabitants as an attractive place in which to live or work or to visit. Many local property developers have become involved in local development agencies and are expected to bring some commercial skills to the process.

2.2 National policy

Taxation

The taxation of land has been an issue of some contention in Britain in the post-1945 era. Partly this reflects a desire to tax those whose wealth is derived from inherited land. A more important issue has been the taxation of gains in land value associated with the giving of planning permission (known as betterment). The last direct tax on gains in land values was the Development Land Tax introduced in 1975 with the Community Land Act. This Act contained a large number of exemptions, which resulted in its yield being so low that the government decided to abandon it in 1986. Today, gains in land values are taxable with other income within company accounts and no special tax regime exists.

Compensation

Generally the issue of compensation for compulsory purchase was not a major issue in the 1980s. Part of the reason for this is that the public sector purchased very little land in this period, but in addition major improvements in the scope and level of compensation were made in the 1970s. Further improvements under the Planning and Compensation Act 1991 introduced the principle that a premium on top of market value could be paid. These seem to have weakened much of the opposition to compulsory purchase on financial grounds, although principled opposition on environmental grounds remains and is possibly increasing.

The environment

Environmental issues were not seen as an important part of the political agenda in Britain until the 1989 European Parliament elections. Since then the government has become more actively involved, especially with such issues as water quality and food purity. In the property field the major change has been the Environmental Assessment requirement for major developments under the EC Directive of 1985, implemented since July 1988. This EC Directive was incorporated into UK procedure in a form that permits and encourages developers to do only the minimum required by the Directive.

Green belts

The green belt policy was inaugurated in the mid-1950s, and has been consistently implemented since. The objective was to stop the spread of urban areas by a policy of containment symbolized by maintaining a green circle of land around the urban area (Elson 1986, Hall et al. 1977). Development was in principle banned from such areas, though a number of exceptions had been made prior to 1979. Not all the land is "green" in the sense that it is farming land or green recreational land. Some of it is derelict, and the example of Stockley Park is a case study of development on such spoiled land (see Ch.6.1).

In the 1980s, pursuing its deregulatory and free market ideology, the government attempted to relax the green belt policy in 1984/5. However, political pressure forced the government to revert to a strong green belt policy in the face of clear evidence that it enjoys popular support among voters. Developers have generally accepted this position (see, for example, the Hertfordshire case study, Ch.9.2) but on occasion have attempted to modify it at both the plan-making and development control stages of the planning process.

The encouragement of owner-occupied housing

The level of owner-occupation of housing in Britain had by 1979 reached well beyond the point where any political party could be critical of the tenure. The Conservative government was not simply non-critical but rather saw the increase in home ownership as an important step in the creation of a centre–right political culture and a property-owning democracy. The increased ownership of housing wealth was seen as important in tying society to the free-enterprise economy. To this end council tenants were given subsidies and the right to buy the house they occupied from the local authority. Council rents were increased to encourage the tenure change.

The land availability process ensured that building land was brought forth and government intervened directly in the setting of house-building expectations in local plans, so that development could proceed in those areas where demand was strong. This was particularly important in parts of South East England where restrictive building policies and high income levels had increasingly made housing very expensive for new entrants to the market.

Social housing

Given the preference for owner-occupation, the rôle of public housing has been seen since the later 1980s as one of providing for people who have special needs (see Ch.1.3). Those who do not have special needs but are not willing or able to become owner-occupiers are expected to use the private renting system, which government attempted to stimulate with several policy initiatives, of which the Falstone Walk case study is an illustration (see Ch.9.3).

2.3 Regional and local policy

Regional guidance

Early in the 1980s the existing structures for regional economic planning were entirely removed, as their continued existence would have been inconsistent with the market orientation of government. However, the diseconomies resulting from uneven regional development and severe regional disparities in land and property values, at their height at the end of the 1980s, have persuaded the government to introduce regional guidance. This takes the form of spatial planning policy guidelines issued by the government on a regional basis, following consultation with local planning authorities and development interests. The next generation of county structure plans and unitary development plans is to be formulated within this guidance.

Transport infrastructure

The issue of transport has been dominated in the post-1979 period by the provision of road space for the nation's transport needs. Even here however the level of investment has not been high and road congestion has increased rapidly. In recent years the government has committed itself to a new road-building programme.

In rail transport the inter-city rail network has been required to become (and now is) profitable, while subsidies exist for lines seen to be socially valuable. Commuter lines are in neither category and are, especially in the South East, severely overloaded and in desperate need of investment. This is not likely to be forthcoming from public sources as the government is pursuing a policy of rail privatization as the solution to the problem. There has been substantial recent interest in urban rail transit in a number of local authorities as part of the solution to urban congestion, and in a few locations these are now under construction or, in the case of Manchester, now in operation. Belatedly, also, there has been recognition of the need to invest in the Channel Tunnel rail link.

The bus system in Britain was highly regulated for much of the post-1945 period. During the 1980s the bus system was deregulated and the ensuing competition did bring prices down on a number of popular routes. The policy did not, however, stop the trend of continually falling passenger numbers that began in the 1950s.

Local economic development

Since the mid-1970s, many local authorities have developed their own programmes of urban economic development designed to attract or retain employment within their area. These programmes often consisted both of physical planning policies (derelict land reclamation, land assembly, industrial improvement areas, construction of nursery and advance factory units and environmental improvements) and of financial incentives directed at small and medium-sized enterprises.

During the 1980s these programmes developed, often with the political objective of counterbalancing the free market thrust of central government with interventionist measures. They often displayed a noticeable degree of complementarity, however, as they increasingly relied on national programmes (regional policy, City Grants, inner-city programme) or the EC structural funds for their financial support. Many cities have become very effective in mobilizing all available national and EC programmes and resources, and developed image-building and marketing strategies (see below). Urban economic development, and the associated competition between cities, takes place at the municipal level in the UK to the extent that it does partly because there is no regional level of government that might otherwise play a rôle.

Derelict land reclamation

This policy existed prior to 1979 when it was commonly used to reclaim land that had been spoiled by coal workings. During the 1980s, urban derelict land was reclaimed with this money and private developers have been permitted to join the local authorities in bidding for it from central government.

Derelict land policy in urban areas has been tied to that of "gap funding". Where development has not been viable, government has in some cases been prepared to fund the gap between the costs of the development and the value of the finished building (see Ch.4.1). This principle underlies Urban Development Corporations, City Grants, English Estates and social housing development.

1 Stockley Park: Business Park, London Borough of Hillingdon (Ch. 6)
2 Newcastle Business Park: Tyne and Wear Urban Development Corporation (Ch. 6)
3 Aberdeen: greenfield housing development on Lower Deeside (Ch. 6)
4 London's office market (Ch. 9)
5 Radlett and Shenley, Hertfordshire, housing market (Ch. 9)
6 Falstone Walk, Fawdon, Newcastle upon Tyne (Ch. 9)

Locations of the case studies

PART II
The urban land market

CHAPTER 3

The framework within which the urban land market functions

3.1 The legal environment

This section concentrates on the legal position in England. The position in Wales is normally identical: the same Acts of Parliament apply, although the rôle of central government is played there by the Welsh Office responsible to the Secretary of State for Wales. In Scotland, which has retained its separate legal system, separate Acts of Parliament are normally required. (The abbreviation sos refers respectively to the Secretary of State for the Environment (the English minister), the Secretary of State for Wales and the Secretary of State for Scotland.)

For example, the 1971 Town and Country Planning Act for England and Wales, which consolidated into one statute all the then current town and country planning (TCP) legislation, is the 1972 Town and Country Planning (Scotland) Act. The 1990 Town and Country Planning Act, described below, was also a consolidating Act for England and Wales; there is as yet no equivalent for Scotland. A consolidating Act is an Act of Parliament that re-enacts in one statute all the provisions of previous legislation that continue to apply, consolidating them in one statute, without introducing new legislative provisions.

English law is a common law system, contrasting not only with Scottish law but also with the legal systems of most other European countries. Another important distinction in relation to several other EC countries is that there is no separate code of administrative law and structure of administrative courts. Furthermore, a fundamental feature of the planning system is that it is based on discretionary control of development rather than on legally binding plans. Again, this contrasts with most other European countries, although not in this case with Scotland. Therefore, the emphasis in the material that follows has to be more on development control than on the development plan system.

The hierarchy of competences

Parliament Parliament is the only legislature for the UK. An Act of Parliament is the most fully considered expression of the will of the government. Parliament cannot bind its successors, and a later statute that is inconsistent with an earlier statute by implication repeals the earlier statute to the extent of any inconsistency. The courts of law, in seeking a basis for a judgment, may deem this implied repeal to have taken place in arriving at a judgment.

In any Act, Parliament may delegate legislative powers to any legal person or body. The Act of Parliament or "parent act" will set out the principles, and the technical details, rules and procedures will be made under the delegated powers. They are then subject to the scrutiny of Parliament the procedure being dictated in the parent act. Approximately 90% of all UK legislation is made in this way. Such delegated legislation comes under many names, including regulations, orders, directions, rules and by-laws, all collectively known as Statutory Instruments.

Legislation in the form of Statutory Instruments made under these powers is subject to the veto of Parliament prior to taking effect. Statutory Instruments have the same legislative effect as an Act of Parliament, but are open to challenge by the courts if the content exceeds those powers. The main TCP Statutory Instruments and their contents are summarized in the Appendix (see Heap 1991, Butterworths 1992).

The rôle of courts of law The courts cannot challenge the legislation of Parliament and may only interpret it. Individual planning decisions can be challenged in the courts not in respect of their planning merits, but only on the basis of a challenge to the legality of the procedures followed.

In addition to the rôle of interpreting legislation, the courts have the rôle of ensuring that the process of decision-making by government officials, local government and statutory undertakers is fair. This is known as judicial review and broadly arises in the following circumstances:
(a) where statutory procedure is not adhered to;
(b) where an official acts outside their statutory authority or fails to carry out a statutory duty;
(c) where the principles of natural justice are not observed, as in the case of a failure to allow a party a hearing. Any party seeking a judicial review must, however, be affected by the decision or actions of the official and be substantially prejudiced by it.

Maladministration Whenever maladministration is alleged to have occurred, reference may be made to the appropriate "Ombudsman" or, to use the correct titles, the Parliamentary Commissioner if it is a central government matter, or the Commissioner for Local Administration if it is a local government matter. Maladministration is defined as failure to follow correct or advertised administrative procedures in such a way as to cause injustice. It is not necessarily an illegal act. The Commissioners have no power to review a decision by competent bodies properly arrived at, however unsatisfactory that decision may seem. Many planning cases are referred to the Local Government Commissioners, and are reported in the *Journal of Planning and Environmental Law* and the annual reports of the Commission for Local Administration, but only a small number are found to be maladministration resulting in injustice.

Planning legislation In England and Wales, the basic legislation is the Town and Country Planning Act 1990, supplemented by the TCP (Listed Buildings and Conservation) Act 1990 (see Chapter 7) and the Planning and Compensation Act 1991. The Acts however give only the framework of the system, detailed procedures, exemptions and rules being made by the SOS under subordinate legislative powers conferred in the Act.

The SOS regularly issues policy statements on planning issues, in the form of Circulars or Policy Guidance notes. Circulars contain advice or instructions to all bodies with legal duties or responsibilities under the TCP (or any other) Acts, and would therefore be directed to Local Planning Authorities (LPAs). Policy statements directed at the whole range of people concerned with planning and development (developers, the public, as well as LPAs) are normally made in the form of Planning Policy Guidance notes (PPGs, see Appendix) or Minerals Policy Guidance notes (MPGs) in relation to the preparation and content of development plans and Circulars in relation to planning control decisions. They guide officials in the use of their powers and advise the public on how planning issues will be dealt with. They may be referred to in any judicial review, although they have no statutory force.

The planning system

For convenience it is best to consider the planning system in two parts: development control and development plans.

Development control The essence of development control (DC) is that development is defined in an all-embracing manner, and that all development requires planning permission.

The 1990 TCP Act repeats the definition of development first enacted in the 1947 TCP Act: section 55(1) of the 1990 Act states that all development requires planning permission and defines development as "the carrying out of building, engineering, mining or other operations in, on, over or under land" and "the making of any material change in the use of any building or other land".

Anyone may make a planning application, whether or not they are the owner or lessee of the land in question: if, however, they are not the freehold owner they must serve notice on those who hold an interest in the land. Applications are made to the district council in whose area the site is situated. The council, as local planning authority (LPA), is required to determine the application within 8 weeks. A period of 16 weeks applies, however, if an environmental assessment under EC Directive EEC/85/337 is required (see below).

Although all development in principle requires planning permission, there are certain exceptions. The SOS has the power to stipulate permitted forms of development or changes of uses where planning permission is deemed to be granted. This power is presently exercised under the General Development Order (GDO), which lists categories of permitted development, and the Use Classes Order (UCO), which provides that change of use of a building within specified classes of use does not constitute development. These are summarized in the Appendix.

Although all applications are addressed to the district LPA, in areas with two-tier local government the upper tier (county in England and Wales, region in Scotland) is notified of all planning applications that raise county policy issues and may notify the district that certain applications must be referred to the county unless they are to be refused. Applications for major housing projects, private new towns, shopping centres on new sites, large industrial complexes, etc. are likely to be county matters. In addition, all applications for minerals operations are county matters.

The SOS may also intervene to "call in" an application for his own decision, normally for proposals of significance for national policy, and may make a Special Development Order for a particular site or area, automatically granting planning permission for the development specified. The operation of the DC system is discussed in detail below.

Development plans Development plans are not legally binding, but they are statutory in the sense that there is a statutory duty to prepare a development plan, and they are prepared and adopted according to procedures laid down in the TCP Acts. They are prepared by LPAs, and different forms operate reflecting the different systems of local government in the counties and the metropolitan areas.

Preparation of development plans may take some years. The first stages involve extensive surveys and consultation with a wide range of interests in the area of the plan, leading to the preparation of a draft plan. Development plans are prepared with considerable input from and consultation with the public. Public participation is a legal requirement and must take place prior to the issuing of a draft plan, and there is a statutory period for objections to be made to the draft plans.

Sustained objections are considered at public inquiries chaired by persons appointed from the planning inspectorate. Following this procedure, any modifications to the draft plan must again be subject to public consultation and the opportunity to register objections. For structure plans, the inquiry is known as an Examination in Public (EIP); for local plans and Unitary Development Plans (UDPs) it is a Public Local Inquiry.

Objections to structure plans are addressed to the SOS, who appoints a panel of three Inspectors to hold the EIP. Unlike a Public Local Inquiry (PLI), objectors may make only verbal representations to the EIP if invited to do so on the basis of their expected contribution to the examination of general policy issues raised. All written objections are considered by the panel, however. Following the report by the panel to the SOS or, since 1992, to the LPA proposing the plan, modifications may be proposed, that may be agreed or a further EIP may be necessary, following which the plan can be adopted.

Development plans are always a material consideration to be taken into account in deciding a planning application. However, the discretionary system operates so that an LPA decides each application on its merits and, if it judges the case to be justified, may disregard its own policy as expressed in the development plan or in other policy statements in deciding the application. This was the case at Stockley Park (see Ch.6.1). This discretion is not unlimited in practice, as any refusal to an application that is consistent with the plan is likely to lead to an appeal, which may be upheld. This discretion does offer the LPA great flexibility, but does not offer certainty to the developer. The only way in which a developer can be certain of the planning position prior to making investment decisions, for example, is by obtaining planning permission.

Development plans in counties The development plan for any area with two tiers of local government has up to the early 1990s normally comprised two "plans", the structure plan prepared by the county council and any local plans prepared normally by the district council. Under the 1990 TCP Act, every district is now required to prepare a plan for the whole of its area. If the proposals of the Local Government Commission (see Ch.1.1) to form unitary councils in place of existing counties and districts go ahead, one consequence will be that the next generation of development plans in these areas will also be unitary and not two-tiered, as at present.

The structure plan is not a plan in the cartographic sense but a written statement of policies and proposals for the county to cover a period of 15 years, illustrated by a key diagram showing spatial relationships, distribution of settlements, communications and areas of restraint. Its aim is to define policies and general proposals for: the use of land and scale of development; the physical environment, environmental protection and improvement;

economic and social development; and transport, communications and infrastructure. The structure plan is accompanied by an explanatory memorandum setting out the reasoned justification for the policies; their relationship to neighbouring areas, and to national and regional policies; and their resource implications. It must reflect national and regional policy or strategic guidance, expressed by the Secretary of State in PPGs. It also provides the framework for the preparation of the more detailed local plans. Every county planning authority, and region in Scotland, is required to prepare a structure plan and to keep it under review. Until 1992, all had to be approved by the appropriate SoS, but this no longer applies. Local plans have not until recently been mandatory in England, though they have been in Scotland. Current government policy is for nationwide coverage of local plans, and the Planning and Compensation Act 1991 has made them mandatory. During the 1980s, local plans were to be prepared only in respect of areas within a district where there was pressure for development or where constraint on development was desired. On this basis, only 14% of district councils had full district coverage by local plans by April 1990. As pointed out above, following the 1990 TCP Act, preparation of a local plan covering the whole of a district council's area is now mandatory under the Planning and Compensation Act 1991. Full coverage of the country is expected by the government to be complete by the end of 1996 (PPG1, DoE 1992). Note, however, that minerals plans in England and Wales are prepared by counties, as are waste disposal plans in England.

Local plans carry forward in more detail structure plan policies and general proposals for a period of 10 years and illustrate on an Ordnance Survey map, usually to a scale of 1:10,000 or 1:25,000 the sites for proposed developments and the boundaries within which particular policies will operate. Their rôle is to offer guidance to developers, potential applicants, and the public on the policy of the LPA for the use of land and development control in its area. Local plans are the only form of statutory site-specific plan. They must conform generally with the structure plan and although the approval of the SoS is not specifically required he has powers to intervene and reject them. Many local plans are being currently prepared. The network of structure plans was mostly adopted in the period 1979–84, and several are now being reviewed.

Local plans are adopted by the LPA preparing them. The LPA appoints an independent Inspector to hold a Public Local Inquiry (PLI) into the plan to hear objections and make recommendations. The LPA considers the Inspector's report and proposes modifications to the plan. A further hearing may be required, or it may be possible to proceed directly to adoption at this stage. The SoS has the power to intervene and direct that changes be made to the plan, or further public participation be carried out, prior to final adoption. Reference is made in the case studies (Chs 6 and 9) to the situation in their respective LPAs. Normally 6–12 months elapse between a PLI or EIP and adoption. PLIs and EIPs themselves last 1–2 weeks.

Where no statutory local plan has been prepared, policy guidance documents or non-statutory plans may be prepared by the LPA. Although these would not have the status of an adopted local plan, they may be accepted as indicating the planning policy of the LPA. (For further discussion, see Bruton & Nicholson 1987, Cullingworth 1988, Greed 1993, Hall 1989, Healey 1983, Healey et al. 1988.)

Development plans in metropolitan districts   Unitary Development Plans (UDPs) were introduced following the abolition of the metropolitan counties and the Greater London Council in 1986 under the Local Government Act 1985. They have two parts: Part I performs the structure plan function and Part II is a local plan. They are prepared by each metropolitan district and London borough council for their whole area and must be approved by the SOS. Very few UDPs have completed all stages: most have been prepared and are undergoing public consultation or Public Local Inquiry, or are awaiting the approval of the SOS. Until a UDP is adopted, the old structure and local plans continue to operate.

Operation of the development control process

Planning permission   In principle, all development as defined in law, i.e. "the carrying out of building, engineering, mining or other operations in, on, over or under land" and "the making of any material change in the use of any building or other land", requires planning permission. The interpretation of these provisions has given rise to much litigation since they were first introduced in 1947. Construction of a building clearly falls within the definition, as do any extension or alteration works. Demolition works were commonly thought to fall outside the definition. However, in 1990 this was countered by a decision of the High Court. The 1991 Planning and Compensation Act has now resolved the situation, where section 13(1) defines demolition as a building operation, that therefore falls within the definition of development (Little 1992).

A change of use may be "material", i.e. significant, and therefore require planning permission even if there is no physical alteration to the structure of any buildings.

The TCP Acts specifically exempt certain operations and uses relating to operations by statutory undertakers and certain agricultural and forestry activities, but the main general exemptions are made by the SOS under subordinate legislative powers conferred by the Act and contained in the

General Development Order 1988 (known as the GDO) and the Town & Country Planning (Use Classes) Order 1987 (known as the UCO). These Orders technically grant planning permission for the operations and uses specified. Further Orders can be made from time to time by the SOS for development within particular locations, for example in Simplified Planning Zones (SPZ) and Enterprise Zones (EZ) (see Ch.3.3).

The exemptions made by the Secretary of State under the GDO are summarized in the Appendix, the most important being modest extensions to dwelling-houses and industrial buildings subject to qualifications relating to height, set back, plot coverage and cubic volume. The Use Classes Order is the more important, deeming planning permission to be granted for changes in use that nevertheless fall within the same category specified in the order. The categories of use are summarized in the Appendix.

A formal application for planning permission may also be unnecessary in the case of an EZ or a SPZ. The Order issued by the SOS establishing each EZ or SPZ grants planning permission for the forms of development specified within the Order. There are very few SPZs but, in essence, an SPZ has the same planning regime as an EZ but none of the financial incentives associated with EZs.

An application for planning permission is made in the first instance to the district council as local planning authority. If the county council considers the subject of the application to be a "county" matter, or the Secretary of State considers it to be a national issue, it may be called in directly for their consideration. In 1990, only just over 100 applications were called in by the SOS. Planning control powers may also be taken over from the district council by an Urban Development Corporation or a National Parks Authority, in which case they would become the LPA.

The decision is made by a committee of councillors (the planning committee) with powers delegated to it by the council, on receipt of the recommendation of the Chief Planning Officer. In the case of some categories of minor applications (often referred to as "householder applications"), in some LPAs the power to make the decision may, by an agreement of the planning committee, be delegated to the Planning Officer and not taken by elected councillors. The decision may be to approve, to approve with conditions, or to refuse planning permission. In the latter two cases, reasons must be given justifying conditions or refusal.

In the case of any decision to refuse or to grant with conditions imposed, or when the LPA has failed to issue a decision within the 8-week period, the applicant may appeal against the decision or failure to issue a decision. This appeal is made by the applicant to the SOS, but is heard and decided in all but a minority of cases by the Planning Inspectorate. It may be heard at a Public Inquiry held by an independent Inspector (known as a Reporter in Scotland), or judged by written representations. In fact, over 90% of appeals are decided by the Inspector and only a small number of cases that are of national significance or are politically highly sensitive are actually decided by the SOS in person. Data on numbers of appeals are shown in Table 3.1.

Table 3.1 Planning appeals in 1989.

% of refused applications taken to appeal	32.1
no. of appeals submitted	28,659
no. of appeals withdrawn	4,257
no. decided by: Inspectors	20,637
Sec of S	424
% of decided appeals allowed	36.7
median handling time (weeks):	
for Public Inquiry	37
written representations	25

Source: DOE Annual Report 1993.

An application can be made either for full planning permission or for outline planning permission. An outline application, if successful, commits the LPA to the form of the development subject to reserved matters stipulated by them such as siting, landscaping or access being approved in a further application for full permission. The option to proceed only with an outline application is useful to a developer in avoiding the expense of submitting a detailed application where there is no certainty of permission being granted. The progress of an application for planning permission is shown in Figure 2.

Unless there is any agreement to the contrary, a planning permission remains valid for five years, after which it lapses unless it is renewed. Renewal in effect requires a new application to be made. Therefore, in the case of outline permission, details must be approved within this period, and, in the case of detailed permission, development must commence within five years. If any change, however minor, is sought in the terms of any planning permission as granted, or the conditions attached thereto, a further application to the LPA must be made and the full process completed again.

In considering any application, the LPA must have regard to the provisions of the development plan, so far as it is material, and any other material consideration. What constitutes a material consideration has been the subject of much judicial comment in judicial reviews of planning decisions. Certainly policy statements, if relevant to the application, are a material consideration, as are environmental issues, infrastructure requirements and economic policy, but all will depend on the particular facts of the application.

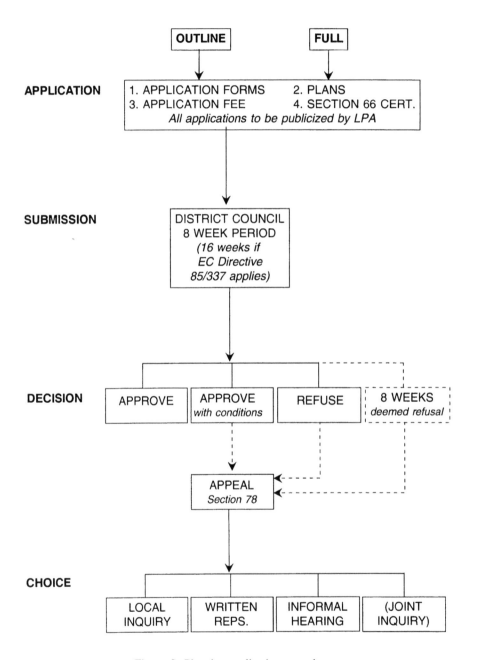

Figure 2 Planning application procedure.

Over 80% of all applications are determined within the 8-week time period. In the case of major and complex proposals, a longer period is normally necessary and the LPA may seek the agreement of the applicant to this effect. During the height of the property boom of the late 1980s, some developers seeking permission for major and complex or controversial proposals put pressure on the LPA by a process known as "twin-tracking". Two applications were submitted: on one, the applicant agreed to any extension of time requested; on the other, the applicant appealed to the SOS against failure to determine the application within the statutory period. Thus a time-clock was set in motion leading to a public inquiry to discourage any delay. This practice has now been stopped by the DoE.

Given the wide definition of development, a considerable volume of planning applications are received each year. Numbers and the times for processing are given in Table 3.2.

The importance of pre-application discussions with planning officials of an LPA cannot be underestimated. Valuable guidance can be given as to what is or is not acceptable to the LPA. During any such discussions the possibility of a voluntary planning agreement under section 106 of the TCP Act may be raised. Under such agreement, planning permission is given by the LPA subject to certain contractual undertakings by the applicant. In this way planning gain is achieved for the LPA that would not be possible by way of conditions attached to the grant of planning permission, that would otherwise be the subject of a successful appeal to the SOS.

Table 3.2 Planning applications. (Number of applications 629,000.)

category	household	change of use	minor	major[1]	other[2]
% of all applications	39.8	8.2	34.9	3.8	13.2
% refused	9.0	27.0	27.0	28.0	na
time taken less than 8 wks	61.0	40.0	39.0	24.0	na

Source: DOE Annual Report 1993.
Notes: na = not available
1, Major application means residential developments of 10 or more dwellings or 1000+ m² of non-residential floor space. 2, Other applications include listed building consent, consent to display advertisements and requests for determination as to the need for planning permission.

Whereas development plans are prepared with considerable public participation, an application for planning permission is generally regarded as a matter between the local planning authority and the applicant. Some LPAs adopt the policy of publicizing all major planning applications in the press, or of always notifying neighbours. However this is not a statutory requirement. Publicity is required only in the following instances where the development:

(a) is a "bad neighbour" development, being a development specified in Article 11 and Schedule 4 of the GDO – places of public resort and buildings exceeding 20 m being important examples;

(b) would in the opinion of the local planning authority affect the character of a Conservation Area or the setting of a listed building under the Listed Building and Conservation Areas Act, section 73;

(c) would in the opinion of the LPA depart from the terms of the development plan and the LPA intend to grant permission.

In all such cases a site notice must be displayed for 7 days and an advertisement placed in a local newspaper specifying that objections can be made within a period not less than 21 days in cases (a) and (b) and 28 day in case (c). In Scotland, notification of a planning application must be served on all occupiers of land coterminous with the development site, being either within 4 m of the site or 90 m of any part of the development in question.

Planning committee meetings are open to the public but the public have no right to speak, objections being considered by the committee on the basis of written objections.

No third party rights of appeal exist in the UK: a member of the public who has been materially prejudiced by a planning decision, and therefore objects to permission being granted, cannot require an appeal to the SOS. This option is open only to the applicant. The only possible courses of action are: (a) referral to the Commissioner for Local Administration on grounds of maladministration (this cannot challenge a planning decision, however unorthodox, if properly arrived at); (b) judicial review of the decision if an abuse of power or procedure can be established, thereby, if successful, quashing the planning decision.

For particular categories of development the LPA must consult with specified bodies, normally government departments or other statutory authorities – for example, the Ministry of Agriculture, Fisheries and Food in the case of the loss of agricultural land exceeding 20 ha; or the Department of Transport where formation of or variation of access to a highway is involved. Consultation may also be required with a government agency such as the Historical Buildings Commission in the case of an application affecting a listed building. Objections received must be considered, and in the case of government departments their will must in certain circumstances prevail.

Instances where such consultations are required are found in Article 15 of the GDO.

Planning permission attaches to the land and its benefit is available to any successive occupiers. Planning permission by way of a section 106 agreement also attaches to the land, the contractual obligations being enforceable by the LPA against any successive proprietor in title who implements the agreed development.

Building regulations In addition to planning permission, building regulation approval is normally required in respect of construction, alteration and extension of buildings and changes of use and the provision of services and fittings. The legislation is contained in the Building Act 1985 and the Building Regulations 1985. The regulations list a schedule of requirements covering structure, fire, site preparation, noise, ventilation, hygiene, drainage and waste disposal, heat, stairways and conservation of fuel and power. The warrant is obtainable directly from the district council, following submission of full plans. Unlike planning permission, building regulation approval is not discretionary and must be given if the plans meet the requirements of the regulations. Furthermore, failure to issue a decision within the time allowed is deemed to constitute approval.

Premises used for public entertainment and the sale of alcohol are also required to be licensed by the district council.

Environmental protection laws Environmental protection is clearly recognized as one of the purposes of the TCP Acts, and in considering an application for planning permission the impact of any development on the environment is a material consideration and may in itself result in a refusal of the application or the attachment of conditions to the grant of permission designed to minimize its impact.

All procedures concerned with controlling environmental pollution are separate from those for applying for planning permission, and are integrated only in cases that fall within the scope of EC Directive EEC/85/337. However, where local authorities are responsible for authorization, they may coordinate the operation of the procedures in conjunction with planning procedures.

It is the intention of the government to establish an Environment Protection Agency to enhance and further integrate environmental protection procedures, but no specific proposals had been brought forward by 1993.

Environmental Impact Assessments The European Community Directive on Environmental Assessments (85/337) was implemented in full in July 1988 by virtue of the Town & Country Planning (Assessment of Environmental

Effects) Regulations 1988 (SI 1199) and the Environmental Assessment (Scotland) Regulations 1988 (SI 1221) (see DOE 1989, Glasson et al. 1994). Environmental Impact Assessments (EIAs) had previously been undertaken voluntarily in cases now covered by the Regulations.

Any planning application for a project that falls within Annex I of the EC Directive (Schedule 1 of the 1988 Regulations), for which an assessment is mandatory, or Annex II of the EC Directive (Schedule 2 of the 1988 Regulations), for which an assessment must be carried out if the project is likely to have significant effects on the environment, must be accompanied by a document containing the assessment as specified in the Regulations. In such cases, the LPA has 16 weeks instead of the normal 8 weeks in which to determine the application.

It is for the local planning authority to determine whether an EIA is required for projects falling within Schedule 2, i.e. those developments referred to in Annex II of the Directive. Guidance is given by the SOS in Circular 15/88. An appeal against a determination requiring an EIA assessment may be made to the SOS. By the end of 1990 approximately 500 EIAS had been prepared and published, of which 60% were under the England and Wales Regulations, 11% under the Scotland Regulations and the remainder under other legislation. Of these, 55 were Annex I Assessments, the remainder being for Annex II projects.

Water The Water Act 1989 established the National Rivers Authority as the main body for the control of water pollution, including coastal stretches, and the management of water resources. Under the Act it is an offence to cause or knowingly permit the pollution of water by any poisonous, noxious or polluting matter or any solid waste. Discharges of polluting agents must be licensed by the Authority and conditions may be attached to any licence. The Authority has powers to prosecute.

Noise The legislation relating to noise also extends to vibration. Neighbour-hood noise is mainly controlled under the Control of Pollution Act 1974 by the environmental health departments of district councils. Local by-laws may however also operate and it is open to any individual through a common law action in the courts to seek an abatement through an injunction and claim for damages. It is for the opinion of the department whether noise constitutes a nuisance, and it has powers to require cessation or restriction of the offend-ing activity or the completion of works to minimize the effect of the activity. A local authority has powers to serve a notice imposing requirements as to the way in which construction work is carried out with a view to minimizing noise. A consent for the methods by which works will be carried out can be obtained in advance from the local authority under the Act.

Air Air pollution is mainly regulated under the Environmental Protection Act 1990 by HM Inspectorate of Pollution (HMIP) in respect of prescribed chemical processes and local authorities in respect of prescribed combustion activities. Affected premises require consent through registration, and conditions may be imposed in relation to construction and use of the premises. More generally, under the Health and Safety at Work Act 1974, there is the obligation imposed on those controlling workplaces to use the best practicable means to prevent the emission into the atmosphere of noxious or offensive substances and to render harmless and inoffensive such substances as may be so emitted.

Health and Safety at Work Act 1974 This Act establishes a comprehensive and integrated legal and administrative framework for securing the health, safety and welfare of persons at work and for protecting other persons against risk to health or safety arising from the activities of persons at work. Inspections of premises to ensure compliance with the terms of the Act are undertaken by and prosecutions are made by the Health and Safety Executive.

Private law relating to land, land transactions and first urban use

Real estate contracts Agreement for the disposal of land is almost always recorded in a written contract prepared in duplicate and subscribed by each party, with each party therefore retaining an enforceable record of it. The contract does not in itself transfer the freehold nor in the case of a lease in excess of 21 years does it create such a lease. The contract merely places the parties to the agreement in the position to demand that the contractual terms be implemented or damages reflecting reasonably foreseeable losses in consequence of the breach paid. Transfer of ownership of property and the creation of a long lease occur only on the registration of the transfer or lease with the land register. Leases for a period of less than 3 years may be by verbal agreement, and in excess of that period but less than 21 years may be created by any written contract between the parties that is subscribed by them and properly witnessed.

If a purchaser is intending to change the use of or to develop land, it is necessary to consider whether, in addition to the need for planning permission, there are any covenants or easements affecting the use of the land.

Covenants A covenant is a form of contract affecting land that is recognized in common law and subject to a degree of statutory regulation. It arises where the owner of land, in agreeing to a sale of part of his land, imposes restrictions and conditions on the transferee's use and enjoyment of the transferred land. Intended to protect the value and amenity of the retained land, they were used extensively prior to the advent of modern planning legislation in 1947. Only a negative obligation may, however, be created under English law. Covenants have commonly been used in the past by developers seeking to preserve a particular character for their housing estate developments, and would normally still be valid. Common examples of covenants are that land shall not be used for industrial purposes, that buildings shall not exceed a certain height or maximum density of development, or that the land shall be used for housing only.

A covenant runs with the land and is enforceable by any successor to the retained land against any successor to the encumbered land. In restricted circumstances it may also be enforced by the owner of land similarly encumbered as in the cases of an estate development.

If a covenant cannot be extinguished or modified by agreement, then the owner of the encumbered land has a right of recourse to the Lands Tribunal under section 84 of the Law of Property Act 1925, as amended. The Tribunal has discretion to extinguish or modify the covenant and make an award of compensation, if appropriate, if any of the grounds specified in the 1925 Act are established by the encumbered proprietor. Grounds specifies in section 84(1) Law of Property Act 1925 as amended by Law of Property Act 1969 section 28 are:

(a) that the covenant has become obsolete, because of changes in the character of the property, or of the neighbourhood or because of other material circumstances

(b) that there has been an express or implied discharge or modification of the covenant by the actions of the parties or their predecessors in title

(c) that the discharge or modification sought would not injure the party seeking to enforce

(d) (i) that the continued existence of the covenant would impede some reasonable use of the land for public or private purposes; and

  (ii) (a) that the covenant does not secure to the enforcer any practical benefits of substantial value or advantage or (b) the covenant is contrary to public interest and

  (iii) that money will be adequate compensation for disadvantage or loss suffered by discharge or modification.

Most titles to land will be affected by covenants. In the majority of cases they will be irrelevant and/or the identity of the party entitled to enforce will have been lost. In most other circumstances the need to resort to the Lands Tribunal can be avoided by payment of a fee for waiver or modification.

Easements Easements are granted to statutory undertakers and public utilities, giving the right of access to their installations, which may be overhead cables or underground pipes, etc., for water, drainage, gas, electricity, etc. These can occur across private land, preventing the erection of any permanent structure that would impede access, as well as on public land such as within the curtilage of the highway.

Subdivision of plots Plots of land may be freely subdivided. It is for the disposer, however, to retain in the transfer such rights as may be necessary in the form of either easements or covenants that are necessary for the reasonable enjoyment of the land.

Instruments for the implementation of plans

Private law Major development projects are sometimes authorized by Private Acts of Parliament, which are subject to a shorter procedure through Parliament. Historically, this was the method adopted in the 19th century by the railway and canal companies. Since the passage of the 1947 TCP Act, this procedure fell into disuse until recent years. It has recently proved an attractive alternative to orthodox planning procedure as a means of expediting authorization of major and controversial projects without the delays caused by major public inquiries of the sort typical of major road proposals in the late 1970s or nuclear power installations in the 1980s. These inquiries could last months or even up to two years. Another feature of this procedure is that it bypasses the requirement for an environmental assessment under the terms of the EC Directive (see above). The best recent example of the use of this procedure is the Channel Tunnel Act 1987 (Holliday et al. 1991).

Public law The principle of replotting land is not recognized in the UK and is not regulated in law. It can be undertaken by private agreement.

It is possible to protect planning proposals by serving a stop notice on a developer, preventing further development activity on site. This may be as the first stage of enforcement procedure or in order to prevent demolition of a listed building.

Compulsory purchase is possible under several Acts (Housing, Transport, Education, etc.) as well as under the TCP Acts. In principle, any Compulsory Purchase Order (CPO) has to be justified in terms of what the land is to be used for afterwards. Existing owners have the right to register objections, which may be heard at a public inquiry by an independent Inspector before any CPO can be confirmed in the original proposed form or modified following the Inspector's report. CPOs under the TCP Acts are rare, as it is easier to justify an Order for a specific purpose. Orders under the Housing Acts may be justified in terms of the quality of the existing dwelling if they are unfit for human habitation, rather than in terms of the after-use. Compensation is normally payable on market value basis. The Planning and Compensation Act 1991 affected this position by introducing for the first time the principle that a premium above market value could be paid in certain circumstances.

There is no form of betterment levy under present legislation, although a long history of attempts to capture betterment for the public purse is a feature of post-war planning legislation (Cherry 1974, 1982, Cullingworth 1988). Improvements do not attract any charge under planning or property law, but in the case of non-residential property may create a higher rateable value on which the Uniform Business Rate is levied. In the case of residential property, improvements could raise the valuation into a higher tax band for Council Tax purposes (see Ch. 3.3).

No offence is committed by the failure to obtain planning permission for a development as defined by the 1990 Act or by the failure to observe conditions attached to a grant of planning permission. The only exception to this is in respect of a listed building (i.e. building listed by the SOS as being of historic or architectural interest).

Compliance with planning control legislation is primarily assured by means of enforcement notices served by the local planning authority under section 172 of the Town and Country Planning Act 1990. An enforcement notice is served on the owner, occupier and all persons having an interest in the offending development. It will and must specify the breach of planning control, the steps required to remedy the breach and the time limit for compliance. A notice comes into effect 28 days following service. Failure to comply with the terms of an enforcement notice is a criminal offence and entitles the local planning authority to enter on to the land to fulfil the requirements of the notice itself at the expense of the proprietor. An enforcement order can go to the extent of requiring demolition of the offending development and reinstatement of the land.

An appeal against an enforcement order can be made to the Secretary of State on any of the grounds specified in section 174 of the 1990 Act. An appeal is in effect a retrospective application for planning permission or a retrospective appeal against conditions imposed on a grant of planning permission. To argue that the development was carried out by a previous occupier is not a valid ground of appeal. However, once a period of four years has elapsed after the date on which the offending development took place, enforcement cannot proceed.

An enforcement order does not in itself stop further construction work or an unauthorized activity constituting development. A developer could, for example, continue with construction in the belief that this would influence the appeal. Cessation of construction or use is achieved by the local planning authority serving a stop notice under section 183 of the 1990 Act in conjunction with the enforcement notice. A local planning authority is, however, liable to a claim for compensation under section 186 of the 1990 Act for losses where an enforcement notice is later quashed on appeal or withdrawn.

Section 91 of the 1990 Act provides that a grant of planning permission is valid for a period of five years. If, during that period, the planning permission is not implemented in full, the LPA can revoke or modify the grant of planning permission by an order under section 97 of the 1990 Act where the development is no longer desirable, having regard to the development plan and other material considerations. A claim for financial compensation for abortive costs and losses under section 107 of the 1990 Act is possible. The order is subject to the confirmation of the Secretary of State unless (a) it is unopposed by the owner, occupier and other parties with a material interest in the land on whom the order must be served and (b) no objections are received following public advertisement of the non-opposition. An objector can require that a hearing is held prior to a decision.

A local planning authority may also serve a discontinuance order under section 102 of the 1990 TCP Act requiring that a use of land be discontinued or buildings or works be removed. The procedure and right of compensation are similar to those for an order revoking or modifying a grant of planning permission.

All of the above notices can be made by the Secretary of State following consultation with the local planning authority. This does not circumvent the need for a hearing for objection.

The land-use proposals given in a development plan and the grant of planning permission will invariably result in an increase in the value of the affected land. Such increases are not the subject of direct taxation, although they may be the subject to a charge to Capital Gains Tax on a subsequent disposal of the land (see Ch.3.3).

The historic concept of the right of ownership is maintained to the extent that an order to construct is not recognized in the UK. A freeholder, and a leaseholder if the terms of the lease permit, is free not to make any beneficial use of his land even where planning permission exists. Instances occur where property developers will leave barren and derelict areas of land vacant in order to place pressure on a local planning authority for a grant of the desired planning permission. A local planning authority could, however, use its wide powers of compulsory purchase in such circumstances but would be required to pay compensation in so doing.

There is no provision in law for the state to have a right of first refusal in respect of a disposal of and interest in land. The only exception to this is in the instance of a disposal by a private sector landlord of residential accommodation within a Housing Action Area. A private developer or the state could, however, acquire the equivalent of a right of first refusal on land by the inclusion of an option-to-purchase clause within the context of a contract to purchase.

Information systems

Public registers of available development land   Under the Local Government, Planning and Land Act 1980, every local authority holds for public inspection a copy of a register of land within the ownership of local authorities and statutory undertakers that in the opinion of the Secretary of State for the Environment is not being used or is not being sufficiently used for the statutory purposes of the land-holding body. At the end of February 1989, 34,000 ha of land was registered. The Secretary of State has statutory powers to direct the sale of any land disclosed on the register where this is required for private development.

In practice, most county or district councils will maintain a listing of land and buildings available for development or occupation by commerce and industry as part of its programme for economic regeneration and development.

Public matters affecting land   Under the Local Land Charges Act 1975, a district council must maintain a register of public matters in which local authorities, government agencies and independent statutory corporations are interested and that bind successive owners of land.

The most important entries are recordings of: decisions on applications for planning permission and conditions attached, if any; listings and designations made under the Planning (Listed Buildings and Conservation Areas) Act 1990 (see Ch. 7); road proposals; designations of areas under public interventionist schemes, such as an EZ or UDC; any planning enforcement proceedings; and any CPO.

The council will provide a list of entries affecting any land and buildings for a fee.

Development control register   In addition, under the TCP Acts, local planning authorities must maintain and make available for public inspection a register of all planning applications, with a plan showing the site concerned and a list recording the type of application, permission sought, decision reached, and dates and reference numbers.

Housing land availability   All county and Scottish region planning authorities used to be required to maintain a schedule of housing land availability, indicating the location of each site on a plan and recording on a schedule the numbers of dwellings for which planning permission may have been anticipated, consistent with the provisions in the relevant structure or local plan policies. This requirement was repealed in 1992, but local authorities are still required to ensure sufficient land is designated for private housing in all new plans.

Register of ownership A public register recording the ownership of freeholds and leasehold interests in excess of 21 years is maintained under the Land Registration Act 1925. The register was introduced as a means of operating a compulsory and simplified procedure known as land registration for the transfer of freeholds and the creation and transfer of long leases. Land registration was introduced gradually, the last registration area becoming operative in December 1990. The register is incomplete, as registration is induced only on a sale of a freehold or the creation or sale of a long lease.

Census A census is undertaken every 10 years under the Census Act 1920, the last census having been taken in April 1991. The information collected in the 1991 Census is detailed in the Appendix and differs only from the 1981 Census in requesting information on ethnic origin. Census results are compiled and published by the government.

Cadastre There is in the UK no cadastral survey, and no interest shown in creating one. Local commercial property taxation is determined by the rental value of the property. Records and assessment are maintained by the Inland Revenue and collected by the local authority.

For private actors who wish to know the ownership of land and interests in land, considerable difficulties can arise. In practice, the information that is required in the market is usually available in the private sector and known by firms of local surveyors (see Ch.4.2).

With the introduction of the Council Tax (to replace the Community Charge) in April 1993, there has been the need to assess the capital value of each house in order to place each dwelling into a Council Tax price band. As the date of valuation was April 1991, the information is seen as almost irrelevant to the current housing market.

Valuation committees There are no valuation committees. Valuation data are not made available by public authorities as a public service. Land and property valuation data are assembled privately by firms of chartered surveyors, etc., as a private resource and are subject to limitations of commercial confidentially. Commercial data services, accessible by private subscription, are also becoming available.

The code of professional ethics generally avoids problems of misleading estimates. Cases of negligence can result in claims against the surveyor's professional indemnity insurance.

Within each local authority there is a District Valuer who has three main rôles;

(a) to estimate the value of land and property that is obtained via Compulsory

Purchase Orders (CPOs) and thereby to set the level of compensation,
(b) to estimate the rent level for each commercial property, such estimates to be used as the basis for the Uniform Business Rate,
(c) to determine which house price band each residential unit should be placed in, this forming the basis for the Council Tax.

In practice CPOs are quite rare. Commercial rent estimates are based on April 1991 comparables, which are increasingly dated, and house price banding is also dated and imprecise. As a consequence, District Valuer data are not relevant to market conditions.

3.2 The financial environment

Who finances and under what conditions: general credit practice
The commercial land and property market The private sector system of property finance is characterized by a highly competitive and de-controlled banking system coupled with a diverse collection of land and property investors. The essential distinction is between equity finance, where the lender takes a share in the future rewards and risks associated with the property investment, and debt finance, where the lender receives only interest payments and where the borrower takes the risks and earns the rewards.

Equity finance via borrowing from the British Stock Exchange is available to property companies that are able to offer potential returns to investors. The borrower is required by Stock Exchange rules to disclose all matters of relevance to investors.

Debt finance is also widely available, though it is normally restricted to a percentage of the asset value so as to reduce the risk undertaken by the lender. Forms of debt vary from simple mortgages to complex discounted bonds where capital growth is the only form of return.

As with any form of commercial loan, the lenders are concerned with risk and returns while borrowers seek to minimize their borrowing costs and any restrictions placed upon them by lenders. The competition to lend, especially for property development, has led to the introduction of increasingly complex loans where issues such as tax efficiency and flexibility have become the dominant criteria.

There are no restrictions on capital import since all exchange controls were abolished in 1980. The complete absence of controls on capital mobility has ensured that the property market is accessible to a range of non-British banks and financial institutions. These have added to the competitive pressures in the market.

Land The issue of land banking is addressed below, but the general rule holds that investment in land is subject to the same credit practices as for property or any other asset. Although land cannot become obsolete, its value can fall and therefore lenders do not treat it differently from other assets. Direct borrowing for the purchase of land usually entails the borrower placing the land as security for the loan.

Property Development The forms of finance for property development have changed considerably in the post-1945 period. During much of the 1980s the property development process was so profitable that funds were made available from a wide variety of sources.

Borrowing from the Stock Exchange was possible up to the stock market crash of 1987. Generally the stock market is attractive to the larger property companies that wish to expand rapidly, but the source of finance is restricted to those periods when new share issues are possible. In theory, funds from this source should always be available for potentially profitable developments, but in practice waves of pessimism may restrict availability. In mid-1993 the Stock Exchange was again becoming a source of property finance.

Borrowing from banks was also important during the 1980s (see Ch.1.4). The banks may lend to the firm or for individual projects where the loan is backed by the company. In each case the bank assesses the risks of the loan and adjusts its charges accordingly. Where risks are very high, banks have been known to raise interest rates to levels that some commentators argue imply that they are sharing in the profits of the development. In most cases, the banks require the developer to put some finance of their own into the project at the outset and a process of monitoring is agreed.

Property investment All investors in commercial property are faced with the problem that returns in the first few years are rarely equal to interest payments on loans used to finance the property. This reverse yield gap is commonly overcome in time by rising rental income and capital gains, but the initial shortfall in returns has ensured that property investment is a risky activity.

There are a wide range of investors in property and each has access to different forms of finance.

○ These have access to both bank and equity finance and during the 1960s and early 1970s were major investors in property. They became less important at the end of the 1970s as the financial institutions, which receive tax benefits upon their investments, became more active. Today they still face competition from the financial institutions but they have regained much of their previous dominance in the market. The property companies' strength is their ability to purchase properties that are in need of improvement, expansion, etc., and after development work they will exploit the improved asset. Their entrepreneurial skill in recognizing development opportunities is their key asset.

○ Pension funds and insurance companies have a steady inflow of funds and a need to find a safe, profitable outlet for them. Part of their investment strategy has involved purchasing property, but they have tended to invest in only the most prestigious properties, particularly office space in London.

Because they are not required to make immediate interest returns on their investments, they are able to ignore the low initial returns from property (i.e. the reverse yield gap), and indeed competition between these organizations has driven capital values up and yields down.

For these "institutional investors", the UK lease structure (see Ch.1.4), with a 25-year term and upward-only rent reviews, makes commercial property an attractive investment medium.

○ There is some investment in commercial property by individuals but it is concentrated in the poorer-quality end of the market. Individual investors are relatively unimportant.

○ Owner-occupiers of commercial property are usually companies and the ownership of property is a commercial decision. They borrow money on the same principles as other borrowings. The lender to the companies (i.e. a bank) may offer lower interest rates for property purchase, especially if the property could be easily resold to another business if the borrower failed to maintain the loan. In recent years many owner-occupiers have sold their property assets and leased them back as a method of raising finance. This process has been encouraged by the financial community.

It must be noted that the British commercial property market is dominated by the leasehold tenure and that the investment rôle of property is seen as highly significant. While the relative importance of the principal investors – property companies and the institutions – has varied over time, the rôle of property as a financial asset and as a means of accumulating wealth has remained. There are many reasons for this, historical, cultural and tax related. What must be emphasized is that property occupiers (tenants) have been subject to the "excessive" burdens of the UK lease and that property is rarely seen as simply a factor of production.

The housing market

Owner-occupation Private housing is usually financed via a mortgage from a bank or building society. While a huge range of forms of borrowing exist, including ECU and foreign currency loans and even life insurance investment loans, the majority of borrowers take out a 25-year flexible-interest loan upon which they make monthly repayments. The interest charges on the first

£30,000 borrowed (for the main residence only) are deductible against the lowest rate of income tax.

Private housing to rent Private housing to rent is often in the hands of owners who have a small number of houses, and their sources of finance are not easy to discern. Modest inherited wealth and profits from small business are likely possibilities. A number of substantial house-owners exist but much, if not all, of their property has been held for a number of years and finance is not a question. Some new landlords have been created under recent legislation that permits short-term tenancies and market rents. This very small sector has been encouraged by a particularly attractive tax regime.

Housing associations Housing associations finance new property development with grants from government via the Housing Corporation. While housing associations were seen by government in the 1980s to be an important element in social housing, they still account for a very small sector of the total market (see Tables 1.11 and 1.13).

Local authority housing Local authorities are able to build social housing, that is normally made available to those who are deemed to be in greatest need in each locality. Central government control is maintained both by restrictions on the spending of local authorities and by limiting grants to local authorities for housing development. During the 1980s the level of local authority house building declined sharply owing to government controls, despite the local authorities' financial position being improved by their receipts from the sale of their housing under the "right to buy" legislation (see Table 1.14), and despite the growth in homelessness.

Credit practice per sector

In the commercial field, the only major sectoral difference is that owner-occupation is more common in the industrial sector than in other sectors.

All forms of finance for development are available in each sector, but in the property investment sphere the best properties (known as prime properties) are often held by the financial institutions. The poorest property, less secure, more risky, etc. is held by those investors who are prepared to take the greater risks.

In the housing sector, the various forms of finance are explained in the previous section.

Land banking

Land banking in Britain means the purchase of land in advance of development activity. Land is commonly purchased before plans that envisage development in this location are created.

It is most common amongst private house-builders, who purchase sites in open countryside at low (agricultural) prices and receive the benefits if and when planning permission is given for development.

With the discretionary planning system that operates in Britain it is theoretically possible that almost any piece of land near an urban area could be given development permission. The absence of legally binding plans means that the development plan status is not a determinant of eventual permission. As a consequence, much of the open land around major cities is owned by property companies, and values are commonly determined by the likelihood of obtaining planning permission in the short, medium or long term (hope value). With sound judgment and good planning advice, much of the land owned in this way is eventually capable of realizing development value.

Research has indicated that profits from land banking form a major element of the profitability of the development industry. Some corporate take-overs have been inspired by the ownership of land banks, and generally the land ownership of land and its potential for development can form and important element in the valuation of any firm.

Transaction costs

Housing It is important to recognize that housing transaction costs can be divided into those that are obligatory and those that are optional.

House buying

Survey of property There are various types of property survey available, from those that are concerned with value to those that examine in detail the structural quality of the building. The costs vary from approximately £50 to £500. This cost is optional to the buyer, but a building society or bank may insist on a survey before a mortgage can be obtained.

Solicitor The use of a solicitor is optional but the buyer will have to pay certain fees and recording dues. Typical costs of fees and dues are £100 plus the payment of Stamp Duty of 1% on properties purchased for over £30,000. The typical solicitor's fees range from £100 to £400.

Finance The lender of finance, where this is required, will usually charge an arrangement fee and require the borrower to take out some form of mortgage protection policy. The typical arrangement fee is £150.

House selling

Surveyors The use of surveyors is optional and not very common.

Solicitors The use of solicitors is optional but the seller will have to pay certain fees. Typical search fees and recording dues are £120, and typical solicitor's fees range from £100 to £400.

Estate agents The use of estate agents is optional and many people do not use them. They charge between 1% and 2.5% of the property's value. Many estate agents are managed by chartered surveyors.

Practice in Scotland differs, especially in respect of the rôle of solicitors (see Ch.6.3).

Applying for planning permission An application for planning permission, whether in full or outline, requires the payment of a fee to the LPA. The scale of fees is contained in the Town & Country Planning (Fees for Applications and Deemed Applications) (Amendment) Regulations 1990. These are summarized as follows:

> *Erection of dwelling-house*
> Outline application £92 for each 0.1 ha to max. of £2300
> Full application £92 for each dwelling-house to max. of £4600
>
> *Erection of other building*
> Outline application £92 for each 0.1 ha to max. of £2300
> Full application £46 where gross area does not exceed 40 m^2 and £92
>   for each 75 m^2 to max. of £4600
> Change of use £92

It is estimated that 50% of the costs of planning administration in LPAs are covered through fees. It is current government policy to move towards full recovery of costs through fees.

Commercial property Fees for commercial property transactions vary enormously and are always negotiated between traders and legal and other professionals.

3.3 Tax and subsidy environment

Taxes affecting the land market
No distinction is made in the UK tax system between land and buildings, the aggregate value of and income from both being assessable for tax.

There are four principal groups of taxes:

(1) income taxation, including Income Tax and Corporation Tax;

(2) capital taxation, including Capital Gains Tax (CGT) and Inheritance Tax (IHT);

(3) consumption taxation, covering Value Added Tax (VAT) and Stamp Duty;

(4) local taxation, covering the Uniform Business Rate (UBR) and the new Council Tax, which replaces the Community Charge.

In summary, the relevant taxes in respect of property interests are as follows:

– Transfer of ownership or leasehold interests: Stamp Duty; Capital Gains Tax for individual transferees; and Corporation Tax for company transferees.

– Grants of leases: Stamp Duty; Income Tax for individuals in receipt of rent and/or premium: Corporation Tax for companies in receipt of rent and/or premium.

– Transfers of property interest by gift or on death: Inheritance Tax.

– Occupation of property: Uniform Business Rate, but in respect only of property occupied for business purposes; Council Tax for occupation of housing.

An overview of the tax system is provided in Table 3.3.

Gains in the value of a property interest are taxable only on the disposal of the interest. Increases in value attributable to development control decisions were directly subject to Development Land Tax until 1986. There are no proposals to reintroduce such a tax.

Local authorities, government agencies, charities, life insurance and pension funds all enjoy a privileged tax position, but these are not discussed in this report.

For partnerships, each member is taxed as an individual on the assumption that income and chargeable gains are shared equally unless the partnership agreement provides otherwise.

The details of the taxes referred to in each of the above categories are dealt with in turn.

Income taxation

Income Tax This is the most important tax in the UK tax system. There is no statutory definition of income beyond the statement that income is taxable if it falls within one or other of the schedules of the Taxes Act 1970. Those schedules are comprehensive and any receipt is in effect subject to the tax if not falling under the head of another tax.

The schedules of the 1970 Act are as follows:

A Rent and other receipts from property

B Income from commercial woodlands

C Interest on certain British securities and securities of foreign

93

Table 3.3 Relevant land and property taxes in Great Britain.

Tax	Taxable person	Object of the tax	Base of the tax	Amount of tax	Tax exemptions and allowances	Recipient	Percentage of the total tax revenue of the recipient
Income tax	individuals	all income under each schedule	annual tax returns	20%, 25% and 40% on higher incomes	none, allowances include individual & couple interest on mortgages and numerous financial investments	central government	27.5%
Corporation tax	companies	trading profit including investment income	audited accounts	25% rising to 33%	none, allowances include capital expenditure on plant and machines and some buildings	central government	7.5%
Capital gains tax	individuals	gains on disposal of an asset	annual tax returns	same as income tax, but treated as top slice of income	first £6000 exempt, allowances include improvement costs and indexation main home excluded	central government	0.8%

Tax	Subject	Basis	Rate	Collection	Exemptions/allowances	Collected by	%
Inheritance tax	individuals	assets at death and gifts made in preceding 7 years	40% at death but progressively reduced for gifts preceding death	return made by administrators of deceased's estate	bequests to spouse + £3,000 per year allowances include proportion of family businesses and agricultural land	central government	0.15%
Value added tax	businesses with turnover >£35,000	final consumption of goods and services	17.5%	quarterly returns to inland revenue	zero rating of food, books etc, and partially exempt eg new residential construction	central government	19.9%
Stamp duty	individuals and companies	transactions, especially land and property	1% of conveyance variable rate for leases	value of transactions being stamped	conveyances less than £30,000 are exempt	central government	0.6%
Uniform business rate	occupiers of commercial property	occupation of property	34.4p in £ in England	full repairing and insuring rental value	periods of non-occupation	local government, but central government grants adjusted	19.5%
Council tax	occupiers of residential property	occupation of property	variable with each local authority but limited by central government	capital value in 1991, property placed in price bands	low income earners exempt single person reduction by 20%	same as above	11.6%

governments and public authorities

D1 Profits of trade

D2 Profits of profession or vocation not dealt with under any other schedule

D3 Interest and annual payments paid without deduction of tax

D4/5 All income other than employment income arising abroad whether or not remitted to UK

D6 Any other income, including income from furnished lets and occasional profits not chargeable to D1 or D2

E Income from employment and pensions, including company benefits

Income Tax is one tax, but each schedule has not only its own rules for computation of the tax but also different rules for payment. The tax year, as for all UK taxes, ends on 5 April in each year.

Until 1984, investment income was taxed more heavily than earned income. Interest, dividends and rent were considered investment income. This distinction has been withdrawn but it made a substantial contribution to the demise of the private rented sector in the pre-1984 period.

The premium received for the grant of a leasehold interest may in certain circumstances be subject to income tax and a complicated formula for calculation of the tax operates by reference to the term of the lease. This arises as an anti-avoidance measure, as a substantial premium might otherwise be sought as an alternative to rent solely to avoid the tax.

Banks and building societies deduct tax at the basic rate on payments of interest, as also do companies in respect of share dividends. A tax credit is received by the individual, and if tax is due at the higher rate then the balance must be paid. The tax paid is not generally recoverable by the individual, except where the individual is a non-taxpayer (e.g. non-working spouses).

Corporation Tax Corporation tax is paid by companies and public corporations but not by partnerships.

Corporation tax is levied on company income from any source and on gains made on the disposal of any company asset. Chargeable income is generally trading profit, including investment income, after deduction of trading expenses, including loan interest charges. Chargeable gains are assessed on the same principle as CGT, with allowances for indexation and roll-over relief.

Dividends are paid to shareholders net of basic rate income tax and this is paid to the Inland Revenue by the company as Advance Corporation Tax, with the balancing payment of assessed Corporation Tax falling due nine months after the end of the company's accounting year. Previously, company profits were taxed prior to payment of the dividend and the dividend was subject to tax again in the hand of its recipient.

For a company trading in property interests, whether a profit is a revenue receipt or a capital gain depends upon the circumstances of the transaction.

Capital taxation.

Capital Gains Tax (CGT) CGT is a tax on gain accruing to an individual on the disposal or part disposal of an asset. A disposal may occur where the asset is sold, exchanged or gifted. It taxes not only speculation but the increase in the real value of an asset during the period of an individual's ownership of it. Indexation relief is built into the computation. For ease of administration, pre-1982 gains are not taxed.

Although assets are widely defined, specifically excluded are a person's home, motor car and movable property disposed of for a consideration of less than £6000. With the annual allowance of £5500, the tax does not normally affect the ordinary individual.

Where the sale of business premises at a profit attracts CGT, but replacement premises are bought, liability to the tax can be deferred until the sale of or cessation of use of the replacement premises for business use. Known as roll-over relief, it is applicable also where replacement land is bought following a compulsory purchase.

An individual's chargeable gain is treated as the top slice of his income and charged at Income Tax rates. Otherwise it is a separate tax. At what time disposal of a category of assets becomes a trade is always a matter of fact and degree.

In conveyancing and leasing transactions, the date of the disposal for the purposes of CGT is not the date of payment of the consideration but the date of the contract for sale or leases becoming unconditional.

The grant of a lease or sublease or the assignation by a landlord or tenant of their interest under a lease for a premium can all be the subject of a charge to CGT. Where the period of a lease does not exceed 50 years, relief to take into account the depreciating nature of the asset is available in addition to that of indexation.

Inheritance Tax (IHT) IHT came into effect on 18 March 1986, replacing Capital Transfer Tax from that date. It is a tax on inherited wealth with a grading of rates (a) to encourage lifetime transfers and (b) to avoid advantage being taken of (a) by a person on their deathbed.

IHT does not affect the ordinary individual, although high house prices in the South East may unintentionally bring more people into its ambit. For the wealthy, careful tax planning to maximize the seven-year chargeable period and reliefs can do much to minimize liability to IHT and allow wealth to be passed on substantially intact to the next generation.

Consumption taxation

Value Added Tax (VAT) VAT is administered and collected by the Board of Customs and Excise. It was introduced in April 1973 as a consequence of UK entry into the EC. Individuals and companies whose annual turnover exceeds £35,000 must register for and levy the tax on goods and services supplied. It is possible to register where turnover does not exceed that figure, and the decision to do so will depend on the benefit of recovering VAT on supplies purchased as against charging VAT to customers on goods and services supplied. A supply is "exempt", "partially exempt", "standard rated" or "zero rated". If a supply is exempt, the supplier does not charge VAT on output to customers and cannot recover any VAT on supplies purchased, i.e. input. A zero-rated supply is one on which the supplier does not charge VAT but may recover VAT paid on supplies purchased.

For the taxable person who is partially exempt, some VAT input tax is recoverable. Where output tax exceeds input tax, the taxable person must remit the difference to the Customs and Excise, or vice versa.

Generally, the construction of commercial and other non-domestic new building is standard rated as also are all conversions, alterations, and extensions. Construction of new dwelling-houses is zero rated. Sales of property and grants of leases are generally exempt supplies. Professional fees, repairs and maintenance and other services rendered in relation to property are generally standard rated.

Stamp Duty Stamp Duty is an ancient tax of some complexity chargeable not on transactions but on the documents that embody them. Payment of the tax is denoted by an impressed stamp on the document. Enforcement of the tax is assured, as chargeable documents that are unstamped are (a) refused registration by the Land Register, (b) not admissible as evidence in civil courts.

A legal right to an interest in land, with a few exceptions, can be created only by a formal written document. Therefore Stamp Duty, although relevant in other circumstances, predominantly affects land-based transactions.

Apportioning part of a consideration in a land-based transaction to movable property or construction costs, in the case of new built premises, is a common method of reducing liability to the tax on the legal principle that delivery itself transfers legal ownership. For the sale of a business as a going concern, good financial advice and skilled drafting of purchase contracts can restrict Stamp Duty liability to that part of the consideration apportioned to land, buildings fixtures, goodwill and intellectual property, i.e. excluding stock and creditors. For long leases of commercial properties, the high levels of Stamp Duty applicable to the annual rent can be avoided by the use of a contractual right to extend the lease exercisable by either party. On the exercise of the right, Stamp Duty is paid as if a new lease were granted.

The requirement to present conveyances and leases in excess of seven years to the Stamp Office within 30 days of completion is an accurate means by which the government assesses the level of property activity.

The tax is administered by the government through the Controller of Stamps at the Stamp Office and is efficient and cheap to collect, non-payment preventing registration of conveyance or lease in the Land Register and use of the documents as evidence in civil court proceedings. Payment is in practice made through a solicitor and this adds to its perception as a transaction cost.

In the residential property market, the payment threshold of £30,000 taken with the interest relief on mortgages up to £30,000 can ease entry for low-income groups onto the housing ladder, but as both thresholds have remained static since 1984 this benefit is restricted to areas of low house prices and discounted council house sales. Any distortion in house prices below £30,000 is minimal.

Local taxation

Uniform Business Rate (UBR)   April 1990 saw important changes in the taxation of non-domestic properties as a source of local government revenue. Properties are valued by reference to their rental values as at 1 April 1988 on a vacant to let basis with full repairing and insuring terms. The tax is then assessed by reference to a multiplier fixed for 1990/91 at 34.8p for England and 36.8p for Wales. For future years, multipliers cannot rise more than the increase in the Retail Price Index. Special arrangements exist for the City of London.

Transitional relief is available to 1994/95, restricting increases by fixed percentages and staggering reductions where properties have benefited. Transitional relief is lost, however, on a change of possession.

Local authorities are responsible for assessing and collecting business rates payable on properties within their authority. They do not, however, retain the rates collected, but pay them into a national pool. The money in the pool was redistributed to local authorities pro rata to the number of Community Charge payers registered in each authority. New methods of distribution have been introduced with the Council Tax.

The Inland Revenue Valuation Office is responsible for compiling and maintaining the new rating list. Appeals against valuations are available in limited circumstances.

Non-domestic properties within a designated Enterprise Zone are exempt from the tax for the 10-year life of the zone.

Council Tax This tax was introduced in April 1993 to replace the discredited Community Charge. The tax is based on the capital value of residential property as if it had been sold at market price on 1 April 1991. Valuation was undertaken by the District Valuer's office, though much of the actual work was contracted out to local firms of surveyors. Each dwelling was placed in one of eight tax bands (A–H). The valuations attached to each of the bands A–H for England are shown, for illustration, in Table 3.4. Each local authority has the power, subject to charge capping or government controls over maximum levels of taxation, to set its own charge for each of the bands. The levels of charge are based on the assumption of two tax-paying adults per dwelling. If the dwelling is occupied by a single tax-payer, a 25% rebate applies.

Table 3.4 Council Tax bands in England.

Band	House price range
A	<£40,000
B	£40,000–£52,000
C	£52,000–£68,000
D	£68,000–£88,000
E	£88,000–£120,000
F	£120,000–£160,000
G	£160,000–£320,000
H	>£320,000

Source: DOE.

Appeals against assessment are possible but, as properties are simply placed in one of the eight price bands, only those coming near a margin may expect to gain much. Some transitional relief has been made available in the first year of its operation so as to restrict the rate of increase of tax liability in comparison with the Community Charge that some individuals will face.

Given the low levels of tax on properties of low value, exemptions for those on housing benefit and students, and the 25% reduction for single-adult households, there has to date been much less resistance to the tax than was the case with the Community Charge.

It is not clear yet how far this tax will affect the operation of the housing market. It may discourage some people from moving upmarket to more expensive property, and there is some evidence that it is encouraging people (often elderly) who occupy high-value dwellings that are larger than their present needs require to seek to move to smaller dwellings coming within lower tax bands. There have also been instances of sellers boasting of a high tax band as evidence that the dwelling is within a high-value neighbourhood.

The Community Charge (Poll Tax) The Community Charge ceased to apply in March 1993. Unofficially known as the Poll Tax, it was the subject of much controversy and criticism after its introduction in Scotland in 1989 and England and Wales in 1990. The central principle was that all residents should pay, either in full or at least 20% of the assessed flat-rate tax even if they were in receipt of state benefits, students or otherwise without income. It replaced domestic rates, which were assessed on the notional rental value of the property as a means of partially funding local government. It proved both difficult and expensive to collect, with serious and increasing shortfalls in receipts due to non-payment. Although the list of charge-payers is officially quite separate from the electoral list and the census, it is suggested that reductions in the registration rates of both, possibly affecting the 1991 national census of population, are attributable to this tax.

Subsidies concerning the land development process

Since coming into power in 1979 the Conservative government has initiated a variety of centrally funded programmes aimed at the renewal of the cities in England and Wales, the majority of which are property based. The current programmes are: Urban Programme, City Grant, Derelict Land Grant, Urban Development Corporations, Enterprise Zones, Garden Festivals, regional grants, regional enterprise grants, regional development agencies, Industrial Improvement Areas, English Estates.

Urban Programme (UP) Established under the Inner Urban Areas Act 1978, the UP operates within 57 urban areas identified by the Secretary of State for the Environment as exhibiting particularly severe problems of urban deprivation. The 57 areas are: Barnsley, Birmingham, Blackburn, Bolton, Bradford, Brent, Bristol, Burnley, Coventry, Derby, Doncaster, Dudley, Gateshead, Greenwich, Hackney, Halton, Hammersmith & Fulham, Haringey, Hartlepool, Islington, Kensington & Chelsea, Kingston upon Hull, Kirklees, Knowsley, Lambeth, Langbaurgh, Leeds, Leicester, Lewisham, Liverpool, Manchester, Middlesborough, Newcastle, Newham, North Tyneside, Nottingham, Oldham, Plymouth, Preston, Rochdale, Rotherham, St Helens, Salford, Sandwell, Sefton, Sheffield, South Tyneside, Southward, Stockton on Tees, Sunderland, Tower Hamlets, Walsall, Wandsworth, Wigan, Wirral, Wolverhampton, The Wrekin.

Priority is given under the UP to projects aimed at reviving the local economy and effecting environmental improvements. Within the 57 areas, local authorities prepare annual Inner Area Programmes identifying projects to deal with economic, environmental, social and housing problems. Approved expenditure by local authorities on these projects receives 75% Exchequer grant through the Department of the Environment. Expenditure under the Urban Programme is analyzed in Table 3.5.

Table 3.5 Analysis of Urban Programme expenditure and outputs, 1988/9-1991/2.

Project type	88/89	89/90	90/91	91/92[1]
Economic objectives				
no. of projects	2,482	3,419	1,995	2,289
expenditure (£m)	96.3	120.5	111.7	122.8
new workshop units	650	699	875	1,159
buildings improved	1,428	1,765	1,271	1,002
new businesses started	526	518	623	1,087
jobs created/preserved	38,840	45,097	29,660	34,242
training places	55,484	61,967	86,528	68,551
Environmental objectives:				
no. of projects	1,338	1,117	1,683	1,381
expenditure (£m)	41.7	44.3	45.0	50.5
buildings improved	6,345	2,381	2,925	8,317
derelict/vacant land improved (ha)	764	743	1,059	1,568
recreational land improved (ha)	1,304	987	392	847
Social objectives:				
no. of projects	4,224	4,126	4,519	3,096
expenditure (£m)	67.2	56.6	61.3	64.8
Housing objectives:				
no. of projects	622	541	760	439
expenditure (£m)	22.8	24.6	25.7	27.6
dwellings benefiting[1]	56,325	78,726	65,928	90,068
land improved (ha[1])	240	217	247	212

Source: DOE Annual Report 1991.
Note: 1, Environmental improvement schemes.

City Grant The City Grant was introduced by the DOE in May 1988 to replace the Urban Development Grant, Urban Regeneration Grant and private sector Derelict Land Grant within the 57 priority areas. It is designed to support projects that are: (a) undertaken by the private sector, (b) capital investments, (c) above £200,000 total project value, (d) unable to proceed because costs, including allowance for a reasonable profit, exceed value, (e) providing jobs, private housing or other benefits.

Projects may be industrial, housing or commercial, but as a matter of policy the grant is restricted to projects within the 57 priority areas referred to above. The grant bridges the gap between costs and profitability to enable the project to proceed. In the first year of operation, 1988/89, £67M was approved by the DoE, in 1989/90 £74M and in the first six months of 1990/91 £31.3M. This represented 194 projects in all, and the total expenditure by the DoE of £172.3M has levered £783.6M in private sector investment.

Derelict Land Grant (DLG) Operating in England under the Derelict Land Act 1982, this grant is available to local authorities, to other public bodies and to voluntary organizations, private companies and individuals, to enable them to reclaim derelict land to bring it back into effective use or to effect an environmental improvement. "Derelict land" is defined by the Act as "land so damaged by industrial or other development that it is incapable of beneficial use without treatment". Land in this context includes buildings that have become so dilapidated or decayed that they are structurally unsound and therefore incapable of beneficial use.

DLG is now incorporated in the City Grant system. Grant is paid gross to local authorities at the rate of 100% of reclamation costs with the increases in value of the reclaimed land being reclaimed by the DoE when disposal occurs. To others grant is paid at the rate of 80% in Assisted Areas and Derelict Land Clearance Areas, and 50% elsewhere, of the net cost of the reclamation works, after taking into account the increase in the value of the land attributable to grant-aided reclamation. The Department of the Environment advises that 40% of DLG in 1989/90 was in respect of inner-city areas. Details of Derelict Land Grant expenditure and derelict land reclaimed are given in Tables 3.6 and 3.7.

Urban Development Corporations (UDCs) There are 10 UDCs in England and one in Wales. They are non-elected bodies, established by the Secretary of State for the Environment under the Local Government, Planning and Land Act 1980, from which Act they derive their powers. Board members are appointed by the Secretary of State and are answerable to him. The UDCs are as follows:

1st generation 1981	2nd generation 1987	3rd generation
London Docklands	Trafford Park	1988/89 (mini UDCs)
Merseyside	Black Country	Central Manchester
	Teesside	Leeds
	Tyne and Wear	Sheffield
	Cardiff Bay	Bristol

Table 3.6 Derelict land grant expenditure, 1985/6–1991/2, (£m).

	1985/6	1986/7	1987/8	1988/9	1989/90	1990/1	1991/2
private sector	2.4	3.4	6.8	11.6	4.3	4.4	8.2
local authorities	71.2	76.8	72	66.5	61.1	69.1	81.5
DLG receipts	(0.5)	(2.1)	(2.3)	(10.4)	(11.4)	(9.2)	(12.4)
Total	73.0	78.1	76.5	67.7	54.0	64.3	77.3

Source: DOE, Derelict Land Grant statistics.

Table 3.7 Derelict land reclaimed, 1985/6–1991/2, (ha).

	1985/6	1986/7	1987/8	1988/9	1989/90	1990/1	1991/2
For development	599	381	588	912	618	627	895
Environmental improvement	461	522	697	575	565	465	627
Total	1,060	903	1,285	1,487	1,183	1,083	1,522

Source: DOE, Derelict Land Grant statistics.

Their overall aim is to secure the physical, economic and social regeneration of their areas with the maximum amount of private investment. In that regard, UDCs reclaim and assemble sites under compulsory purchase powers; assist with the provision of transport and other infrastructure; improve the environment and help provide improved social facilities; and offer financial assistance to private sector development. Overall they finance the gap between the costs of development of certain sites and the value of the finished developments.

UDCs are short-life bodies. The first generation are expected to complete their work in 10–15 years, the second generation in about 10 years, and the third generation in about 5–7 years.

Table 3.8 details English UDC achievements to March 1990. In attracting £9.31 billion private expenditure there has been public expenditure of £1.25 billion, a leverage ratio of 7.4:1.

Table 3.8 English UDC achievements to March 1992.

	land reclaimed (ha)	homes completed	industrial/ commercial development ('000m²)	infrastructure built or improved (km)	permanent jobs gained	private sector investment (£m)
London Docklands	582.6	16,090	2,020.0	84.2	26,000	9,100
Merseyside	337.3	1,367	342.2	44.5	2,000	199
Black Country	119.0	1,289	402.9		3,000	246
Teeside	280.3	315	176.6	14.0	6,025	580
Trafford Park	87.3	70	312.6	7.9	8,231	556
Tyne & Wear	336.9	1,400	385.2	0	9,237	354
Bristol	12.0	110	30.0	0	4,483	44
Central Manchester	19.2	780	75.9	0	750	187
Leeds	37.3	180	255.0	4.2	3,875	116
Sheffield[1]	76.7	0	224.7	4.7	5,909	484
Total	1,888.6	21,601	4,225.1	159.5	69,510	11,866

Source: DOE Annual Report 1991.
Note: 1, Includes complex commenced prior to establishment of UDC.

Enterprise Zones (EZs) There are 24 EZs in mainland UK, of which 18 are in England. Zones are designated by the Secretary of State for the Environment under the Local Government, Planning and Land Act 1980, following the preparation of a scheme for the zone by the relevant local authority at the invitation of the SOS. Each zone has a life of 10 years and they vary in size from 50 ha to over 450 ha. Businesses located within an EZ receive the following benefits during its life:

(a) They are exempt from non-domestic rates (now the UBR). The cost of rate revenue forgone up to the end of March 1990 has been of the order of £270M.

(b) 100% of construction costs, but not the cost of land, of industrial and commercial buildings may be set against Income or Corporation Tax, either in the first year of trading or on a straight-line basis over a period specified to the Inland Revenue. A developer leasing units within an EZ may therefore set construction costs against rental income, or pass the relief to a purchasing freeholder.

(c) Development proposals in accordance with the classes of development specified in the scheme for the EZ are deemed to have planning permission automatically. The normal discretionary system applies only to development other than that specified in the scheme. Any automatic planning permission so granted will however be subject to such conditions or limitations as are specified in the EZ scheme.

Current government policy is to designate further EZs only in exceptional circumstances. They have been widely criticized as expensive and encouraging "fence hopping" by local businesses, as opposed to attracting new businesses to the locality. The most recent example of EZ designation is in Sunderland in 1989 following closure of the shipbuilding industry there, and it has been suggested that an EZ may be designated for North Shields and Wallsend if the Swan Hunter Shipyard closes.

National Garden Festivals Garden Festivals have been held in Liverpool 1984, Stoke on Trent 1986, Glasgow 1988, Gateshead 1990 and Ebbw Vale 1992. Located on inner-city derelict land, the aims of the festivals included accelerated reclamation of derelict land, short-term economic gains and improved image for the host city. On closure of the festivals,, which run for about six months, the land is released for development for housing, leisure and commercial uses. The majority of the costs of the festivals were met by the DoE, funding being channelled through the UDCs, Urban Programme and DLG, with contributions being made by the relevant local authorities and the private sector. The first three English festivals resulted in the reclamation of around 240 hectares of derelict inner-city land. There are currently no plans to continue with Garden Festivals.

Regional development grants Regional development grants were available from the Department of Trade and Industry from April 1972 until 12 January 1988 in respect of approved capital expenditure incurred in providing new buildings and works and on adaptations to existing buildings, and in providing new plant and machinery, on qualifying premises. Qualifying premises were those used wholly or mainly for industrial activities or research or training in relation to industry. This definition was extended in 1984 to include certain service sector activities, including certain computer, mail order and credit card services. Expenditure on site clearance and other preparatory works was eligible for assistance, but not the cost of the land. Payment of the grant was restricted to designated development areas and payable at the rate of 15% of approved capital expenditure, with the alternative of £3000 for each job created being introduced in 1984. Although in effect ceasing on 12 January 1988, the grant continued to be paid in respect of applications received prior to that date. In the year ending 31 March 1990, £203.9M was paid in grant assistance.

Regional enterprise grants These were introduced on 1 April 1988 as a replacement scheme to regional development grants. They apply, however, only to businesses employing fewer than 25 people in development areas. Assistance is available to both investment and innovation projects. In relation to investment, eligible activities include most manufacturing and some service sector projects. Assistance is available at 15% of expenditure on eligible fixed assets in the project up to a maximum grant of £15,000. Eligible costs include plant and machinery, buildings, purchase of land and site preparation. In relation to innovation, a grant of 50% of eligible costs up to a maximum grant of £25,000 may be given to projects that lead to the development and introduction of new products and processes. All costs up to the point of commercial production, including capital costs, may be assisted.

Regional development agencies A number of these have been established either as a result of government initiative (especially in Scotland and Wales, e.g. the Mid-Wales Development Corporation) or as a result of local authority and other regional initiative. An example of the latter is the Northern Development Company, based in Newcastle upon Tyne, covering the northern region, which is a regional promotion agency set up on the initiative of local authorities, trades unions and local business in 1987.

Highlands Enterprise (formally the Highlands and Islands Development Board) was established in 1965 to address the problems of the depopulation of the Highlands and Islands of Scotland. This government agency had an annual budget in 1990/91 of £33M and its board members are appointed by and answerable to the Secretary of State for Scotland. It is charged with the economic and social growth of its predominantly rural area, in addition to promoting training and skills. It provides commercial premises for rent and sale and gives financial assistance to a variety of agricultural, tourist, commercial and industrial projects through grants, soft interest loans and equity investment.

Scottish Enterprise (formally the Scottish Development Agency), which was established in 1975 and whose board members are also appointed by and answerable to the Secretary of State for Scotland, had an annual budget for 1990/91 of £97M. It operates throughout Scotland to promote economic development, enhance skills training and improve the environment, but restricts those activities within the area of Highland enterprise to major projects.

An agency exists in Wales with similar powers and responsibilities to Scottish Enterprise.

Industrial Improvement Areas   The concept of an Industrial Improvement Area (IIA) was developed by local authorities in Tyne and Wear and in Rochdale. It has become a widely adopted feature of local economic development and inner-city policies.

The essence of an IIA is that an area, usually in an inner-city location, is designated to indicate that it will remain in industrial use for a long period (say 30 years). It may be an area of existing industry or land on which industrial development is proposed. IIAs that are already developed typically contain small firms, often in cramped premises or with poor access. IIA status is given: to indicate that local authority grants and loans for improvements to plant, machinery, premises, access, etc. are available; to generate business confidence and create conditions in which bank loans may be offered; and to focus environmental improvements.   IIAs may also be designated where land reclamation is taking place or land is being made available for new factory premises. These would normally be in the form of small advance factories or nursery factory units ranging from $50 \, m^2$ to $500 \, m^2$ of industrial floorspace.

IIAs were first designated under local legislation, the Tyne and Wear Act 1976, and later under the Inner Urban Areas Act 1978, in which the concept was extended to include commercial as well as industrial areas. The terms CIA (commercial) and SIA (shopping) improvement areas are sometimes used.

English Estates   This agency has its origins in the period of mass unemployment in the 1930s. It is charged with providing industrial space in areas of high unemployment as a means of attracting mobile firms to specific localities. It has been actively involved in developing programmes of advance factories, starter units, managed workshops and reclamation of sites. It will either sell or rent premises to occupiers and uses its capital grant from the Department of Industry to produce premises at a price below the costs of production.

In many areas, the long-standing rôle of English Estates has ensured that private developers are unable to compete. Almost all development activity is therefore undertaken by English Estates, though the property investment market is still in private hands.

During the 1980s, the future rôle of English Estates was widely debated and it was generally encouraged to restrict its rôle to development where there was very little chance of private activity.

Currently English Estates manages 6909 units comprising a total area of $1,998,443 \, m^2$. During 1989/90 it built $115,768 \, m^2$ of industrial space.

The future The government has created a new organization, the Urban Regeneration Agency (URA). It has taken over the activities funded by the Derelict Land Grant and City Grant and will also undertake the work of promoting the provision of industrial space, previously the responsibility of English Estates.

The primary aim of the URA will be to "enable vacant and derelict land in urban areas to be brought into use". It will operate primarily as an enabling body, moving on to new tasks once its work is complete and the private sector is able to take over the project. It will have powers of land assembly via compulsory purchase and vesting from other public bodies, and it can use grant aid to finance reclamation and development.

The URA should begin work before the end of 1993. Its future significance will be determined by both its funding level and its method of operation.

Subsidies affecting the housing market

At the outset it is important to recognize that subsidies into the housing sector have always been controlled by central government, and local authorities have had very little discretion in their application. This is consistent with the tradition in Britain that local authorities are subservient to central government.

Subsidies to owner-occupation The owner-occupied sector receives two major subsidies. First, if the buyer borrows to finance the purchase of a home, then the interest on the first £30,000 is deductible against income tax. There has been some debate in Britain as to whether the tax subsidy should be eliminated, but there is considerable public support for it. There is also a debate about whether the tax subsidy actually helps people buy property, because it is believed that it raises demand and prices in many locations.

The second subsidy is that, unlike most other capital investment open to the individual, the capital gains on a person's first house are not taxed. Furthermore, the level of tax on capital on death starts at a level well above the typical house price in most areas.

Finally, it should be noted, that while inflation is not a subsidy, there is little doubt that the individual can almost fix their house purchase costs in money terms and allow inflation to reduce the real cost. For the rent payer there is the probability that rents will rise with inflation over their lifetime.

Subsidies to local authority housing Local authority finance is both complex and subject to frequent changes. The housing owned by local authorities is financed via its Housing Revenue Account. This account, which must balance, has an inflow of money from rents and direct government subsidy and an outflow of debt repayments and administration and repair costs. The level of debt and hence debt repayments is controlled directly by central government and therefore any increase in government subsidy will have the almost direct effect of reducing rent levels.

Since 1979 there has been a general reduction in the real level of government grant, and consequently local authorities have been forced to raise house rents to balance the Housing Revenue Account. Rents for individual houses are set relative to one another, so that the historic cost of a new house does not determine the rent level of the individual dwelling, i.e. there is a process of rent pooling.

Subsidies to housing associations Housing associations have expanded rapidly in the last 10 years, but they still form a small proportion of the rented sector. They receive a capital subsidy from central government for approved investment so that they can charge rents that are competitive with, but not below, local authority rents. The government in effect controls the development of housing associations via its control over this subsidy.

Subsidies to the privately rented sector One of the reasons for the decline in this sector has been the absence of subsidies. The current government has wished to encourage the option of privately rented accommodation as an alternative for those who could not, or chose not to, enter the state systems. To achieve this it has created the Business Expansion Scheme. Investments of up to £40,000 per annum by an individual in the ordinary share capital of a company letting residential property on assured tenancies attract income tax relief at the higher rate of tax. Further, the shareholding is exempt from Capital Gains Tax on its first sale. This scheme ran until the end of 1993. In the first year up to 4 April 1989, 1934 predominantly new companies attracted investment of £362M under the scheme. Of that total investment, £298M was in London and the South East of England. The Business Expansion Scheme has not affected the process of decline of this sector.

Improvement grants Grants for the improvement of houses to ensure that each home contained basic amenities such as inside toilets were readily available in the 1970s and today relatively few houses lack these amenities. In practice, the funds went mainly towards the improvement of older housing, which was intended to benefit the poorer members of the community but often led to gentrification.

Improvement grants are still available but they are now means tested under the Local Government and Housing Act, and the rules ensure that only the very poor, in the very worst housing are eligible for assistance.

Housing Benefit Those who are poor receive benefits towards their housing costs. For poor owner-occupiers, such as those newly unemployed, there is temporary help towards the interest cost of their mortgage, but in practice many who suddenly find themselves poor are forced to sell their home and find another tenure. If this is not possible, and in 1990–93 the fall in house prices made the selling of houses very difficult, the mortgage lender will re-possess the house and if necessary force the owner into personal bankruptcy.

For all renters there is a system of Housing Benefit that gives support on a sliding scale depending on income and family size. There are upper limits to the level of benefit and this limits the rent which the tenant can pay. It is quite common for the government to reduce its subsidy to the Housing Revenue Account, which forces the level of local authority rents up and at the same time triggers additional payments of Housing Benefit. Of course, those who are not receiving Housing Benefit have to pay the full increase in rent and are therefore in effect encouraged either to buy their own council house or to move to owner-occupation directly.

Discounts for the right to buy Council tenants are now able to purchase the home they occupy under the current "right to buy" scheme. Discounts rise from 30% to over 60% for some hard to sell properties. The local authority receives the receipts from the sale of its property and these normally exceed the historical debt incurred in building the house because inflation in Britain has raised house prices well above their historical building costs.

Current trends During the last 10 years the main changes in the system have been that Housing Benefit and mortgage tax relief have become the major housing subsidies, while the local authority non-means-tested subsidy to tenants has declined dramatically. The policies have encouraged the growth of owner-occupation and directed subsidies towards only the poorest in the community.

Subsidies affecting the commercial market

Depreciation in the form of capital allowances set against taxable income is available within an Enterprise Zone and also in the following circumstances under the Capital Allowances Act 1968, as amended:

(1) for construction costs and expenses, but not land costs, of hotels, industrial buildings and buildings used in connection with forestry and agricultural activities; depreciation is on a straightline basis at 4%;

(2) for plant and machinery on a reducing balance basis at 25%.

What constitutes industrial buildings is elaborately defined. The general effect of the definition is to confine allowances to productive as opposed to distributive industries. A non-qualifying use of up to 25% of the industrial building, such as offices or showroom, will be ignored for the purpose of the allowance, but if in excess the allowance will be reduced in proportion to the non-qualifying use.

Capital allowances can be set against rental income or transferred to a purchaser or tenant under a lease exceeding 50 years. A balancing allowance will, however, apply where the written-down value is greater or less than the consideration attributable to the asset.

CHAPTER 4
Prices

4.1 Price setting

In the British commercial property market, capital values and rents are essentially determined in an open and uncontrolled market. Public sector intervention does occur through the control of land supply for particular uses via the planning system and through subsidies to development in certain locations. Generally, however the price of what is available to be purchased is outside the control of government. The owner-occupied housing sector is similar in that prices are determined in uncontrolled markets.

The simplified model of the steps in the process can be considered as follows:
(a) farm land – agricultural value;
(b) farm land with some indication of the possibility of planning permission for development (not green belt, or possibly in structure plan) – hope value;
(c) farm land with site-specific allocation in local plan – higher hope value;
(d) farm land with outline planning permission, and therefore certainty for the developer – development value, allowing for development costs;
(e) farm land with detailed planning permission – value based on expected market value of development;
(f) developed site – market value.

Land values
The value of undeveloped or derelict land is determined by the profitability of the development that can be undertaken on the land. The technical practice is known as "residual land valuation". For each proposed development, the full costs of the development, (including the developer's profit) are estimated and the sum is subtracted from an estimate of the value of the finished building. The resulting figure (the residual) is the maximum that the developer would be willing to pay for the land.

The estimates may be made more certain in the commercial sector by the use of fixed-price building contracts and agreements to sell the completed development to an investor at a fixed price.

The actual value of the land is determined by competitive bids and, as each developer can undertake the same calculation, the outcome is often not far short of the maximum value. It should be noted that the seller can also do the same calculation and hence will be aware of the value of the land. The extensive use of independent specialist advisers (surveyors) by all parties to the exchange ensures that mis-pricing is minimized. The discretionary nature of the planning system does, however, create uncertainty.

The government does not at present generally intervene in the process of land assembly, though in the post-1945 period there have been a number of attempts to change this position. In particular, the Community Land Act of 1975 was intended to bring government to the centre of the land assembly processes. It was fiercely contested by both land-owners and local authorities, who objected to central government having a dominant rôle in the selection of land for development.

During the 1980s the only major public sector rôle in land assembly was that of the Urban Development Corporations, which have purchased land in accordance with the compulsory purchase regulations. The government has also been actively trying to reduce the public sector's total land holdings. Some development of the state's rôle may occur with the creation of the Urban Regeneration Agency (URA).

Commercial property

Capital values The value of commercial property is determined in a free and open market. The purchaser of freehold property as an investment is in effect buying the right to a stream of future rents from the occupant. If the property is newly let on a 25-year lease to a reputable company or a government department then the purchaser will have considerable certainty as to the security of the income, for the only way in which the tenant can avoid payment of the rent is via their bankruptcy. Furthermore, if the property has modern facilities and is in a popular location then it could be re-let at a later date if necessary. If these characteristics do not exist then the value of the investment is suitably adjusted.

In all circumstances, however, the purchaser cannot know the future level of rents, though the lease will normally contain a clause that restricts rents from moving downwards.

The value of a property is commonly calculated as a multiple of the full rental value of the property. The common comparative term is the yield, which is the reciprocal of the multiplier. For example:

Full rental income	=	£100,000
Current multiple	=	20
Capital Value	=	£2,000,000
Current yield	=	5%

If the above transaction were to take place then it would provide evidence that the current yield is 5% and if many transactions took place at this yield then the market yield would be said to be 5%. Valuers would then use this market average to estimate the value of all similar properties, though they would make adjustments as indicated above.

The current yield is related to capital market interest rates because lending on fixed-term rates is an alternative to investment in property. If interest rates rise or fall, then the yield will tend to rise or fall though, as property is a long-term investment, it is not susceptible to short-term movements in interest rates.

The major determinant of property values and hence yields is the expectation of rental growth. When rents are expected to rise, investors will be willing to bid the price of property up (and therefore yields down). In these circumstances a yield of 5%, well below the 1980s rates of interest, is quite common. When expectations of rental growth decline, then demand for property will fall and prices will also fall. The yield will rise. An example of this process is given in the London office market (Ch.9.1) and evidence of changing yields is provided in Chapter 1.4.

Within a geographical region yields can also differ if investors believe one location is more likely to experience rental growth than another.

Rents   Rents are determined in accordance with current local market conditions. The conventional terms of the lease are outlined in Chapter 1.1 and reflect the fact that in Britain the market has been dominated until recently by investors, who have consistently been able to force their conditions on tenants.

Land values and the development decision   A developer buying land with full planning permission would expect to pay the full residual value. In practice, bids for the land would be sought by the land-owner and the residual value would be determined by market competition. The full residual value would be the maximum price the developer would be willing to pay for the land and, if the site needed to be improved before work could begin, then the site improvement costs would be deducted from this sum.

It is important to recognize that residual value calculations are understood by developers, surveyors, town planners and the government. They are used to calculate the degree of support that development projects receive under the

City Grant and English Estates, development corporation, and development agency subsidies. The case studies of the Newcastle Business Park (Ch.6.2) and Falstone Walk (Ch.9.3) illustrate this approach.

One complication to this process is that the valuations are commonly undertaken before planning permission is given, and buyers therefore have to predict the type of planning permission they will eventually be given. Plans may give some guidance on density of development and land-uses, but the discretionary nature of the planning system does create some uncertainty.

Generally developers may avoid paying the full residual land value in a number of ways. Land may be purchased in advance of application for development. The purchaser will have to pay sufficient to attract the land from its current use, and in a competitive market the value will be determined by the expectation of obtaining planning permission in the near future. Land is said to have hope value and it is held as a "land bank" (see Ch.3.2) by the developer. This approach is particularly common in the housing sector, where house-builders usually hold more than five years' supply of land. Not only is it a profitable process but it permits the builder to adjust production more easily to meet variable demand levels (as the case study in Ch.6.3 illustrates). The process is reinforced by the effective low rate of tax on the increase in land value due to the receipt of planning permission (i.e. betterment).

Another approach is to purchase an option to buy land and then to apply for planning permission. If permission is obtained, the option is taken up at the price defined in the option. If permission is not obtained, then the only loss to the developer is the option purchase. This is a popular approach where the development application is highly speculative.

A further approach used in urban areas is to assemble sites over time, paying the price of the current use value while attempting to ensure that the current users are not aware that a site is being assembled for a higher-order land-use. It is common to use a number of companies to obtain the land and to move the ownership to one company only when the site is fully owned. In practice, this process has been the cause of some urban dereliction. Furthermore, some companies have stooped to illegal methods to "persuade" owners to sell their properties.

A final approach in an area where planning permission is unlikely to be given is to purchase the site at the current use price and then to persuade the local authority to give planning permission in exchange for benefits to the local community (or planning gain, see Ch.4.3).

Overall, it can be seen that developers will attempt to avoid paying the full development value but in urban areas it is often unavoidable.

Again it must be emphasized that the property market is dominated by the investment market. Land values are commonly determined by the profitability of development, which is itself strongly influenced by the investment valuation of the completed property. Land investment is seen as just one further form of property investment, and increasing land values are normally seen as legitimate returns rather than speculative gains.

Owner-occupation of commercial property The capital values that owner-occupiers of commercial property pay are the same as those determined in the investment market. For a property seller a sale to either investor or owner-occupier is equally acceptable. In the office and retail sectors the investment market is clearly dominant, but in the industrial sector owner-occupiers form an important element of the market and are therefore influential in the determination of market value.

Housing

House prices are determined by private agreements between seller and buyer. They therefore reflect market forces. Demand in the long term is influenced by population and household formation. In the short run, demand is heavily influenced by financial conditions because the majority of purchases are funded by borrowing from banks or building societies. Since the deregulation of the banking system in 1979, the mortgage market no longer suffers from quantitative restrictions on lending. Today the only restriction is the cost of borrowing (the mortgage rate), which varies with other rates in the economy.

The demand for housing in Britain tends to be cyclical. The most recent boom in the late 1980s saw house prices rise to high levels, not only absolutely but also in comparison with income levels. While the average house price to income ratio has been about 3.5, the figure in 1989 was 4.56 (see Table 1.15).

At the local level, it is common for the house seller to advertise the house at one price and then negotiate with prospective purchasers (see case studies in Chs 6.3 and 9.2: note that practice in Scotland differs from that in England). Prices may be adjusted up or down in the light of offers, and in boom periods houses may be sold by auction. Where the house is of a standard form and values can be estimated from adjacent properties, neither sellers nor buyers require the services of a surveyor and there is no legal requirement to have one. For more individualistic properties, a surveyor is necessary as direct comparisons are more difficult. In practice, sellers in England and Wales usually, though not always, appoint an estate agent, whose rôle tends to be one of agent rather than valuer. This rôle is traditionally taken by solicitors in Scotland. Many of those working as estate agents do not hold RICS qualifications and do not offer the full range of professional services that a chartered surveyor would (see Ch.4.2).

117

Houses are advertised on the basis of facilities, location, number of habitable rooms and quality of design and maintenance. Cost per square metre calculations are rarely used.

Housing development The valuation of land for housing development is in principle the same as for commercial development, and the residual land valuation will vary with regional house price levels. As was noted in Chapter 3.2, the process of land banking is important in the housing sector and therefore the full value is often not paid to agricultural land-owners.

In housing it is also possible to find locations where the market value of housing to the purchaser is less than the cost of the building and in these locations development would not take place unless government subsidies were available (see City Grant, Ch.3.3). Public and private sector house rents are explained in Chapter 1.4.

4.2 The actors and their behaviour

Actors in the process

Buyers, sellers, landlords and tenants can all act independently within the property market and are not required by law to use professionally qualified people. In practice, in the housing market sellers commonly use the services of an estate agent, who is often a chartered surveyor though there is no legal restriction on entry to the estate agent occupation. Buyers may seek advice about building structure and services from a building surveyor, and both parties almost always use a solicitor, though once again there is no legal requirement for this. Generally the owner-occupier housing sector is heavily influenced by the surveying profession in their rôle as estate agents, but much of the exchange activity takes place without professional advice.

In the commercial property sector, virtually all property exchanges are handled by surveyors, who provide independent advice to "their side" during negotiations over prices and other matters. Where a lease specifies a rent review, both landlord and tenant seek advice from surveyors, and where agreement cannot be reached a tribunal of surveyor(s) is required to assess a correct figure.

The land assembly and property development processes are predominantly undertaken in the private sector and again surveyors are commonly employed to give professional advice at each stage in the process. Where the public sector intervenes, such as with land assembly by an Urban Development Corporation, it is noticeable that the Corporations employ surveyors to give them advice about the local land and property market.

It can be seen therefore that the surveying profession is the key actor in the setting of land and property prices. As a profession, surveyors share methods and approaches to valuation issues and share a language and training. Seen in a British perspective, they act as a lubricant to property market processes and are deemed irreplaceable.

In order systematically to review the rôle of all the actors in the planning process and urban land and property markets, they are considered in four categories: public actors, private actors, professions and pressure groups.

Public Actors Within the state system there are a number of actors who have a rôle in the land and property system.

Central government The government department centrally responsible is the Department of the Environment (DOE), headed by a member of the Cabinet, the Secretary of State for the Environment (SOS). It has responsibility for strategic planning advice, development control and planning appeals, the Inspectorate, environmental protection, historic buildings, regional assistance, inner-city policy, City Grants and the Urban Programme, local taxation and public housing.

The equivalent powers are exercised in Scotland and Wales by the Scottish Development Department and the Welsh Office, each with its own SOS in Cabinet. In the case of Northern Ireland, all planning powers including those of LPAS are exercised by the DOE(NI), a branch of the Northern Ireland Office, responsible to the SOS for Northern Ireland.

Local authorities They are subject to the legal rule of *ultra vires*, which means that a local authority may undertake only those tasks for which it has been given the statutory power or legal competence, unlike private persons whose actions are legal in the absence of any law against them. Local authorities carry out the duties assigned to them by law or by direction of the SOS. All local authorities with planning powers are referred to as local planning authorities or LPAS (see Ch.1.1 and Fig. 1).

Development corporations This form of planning authority was first established in the form of New Town Development Corporations established under the New Towns Act of 1946. Most of these have now been abolished, or shortly will be. The major examples now are the Urban Development Corporations, created in the 1980s. They are non-elected planning authorities directly appointed by, and responsible to, the SOS. They have powers of compulsory purchase and can make extensive use of public funds to subsidize the process of property development in areas where commercial activity is perceived as unprofitable (see Ch.3.3).

In many towns the local authority has created its own development corporation as an independent agency charged with encouraging redevelopment of the locality. These bodies have limited resources and are often purely promotional agencies.

Housing Action Trusts Housing Action Trusts (HATs) were introduced under the Housing Act 1988 in order to take over the assets and management of council housing estates that were in need of comprehensive reinvestment. However, a provision in the legislation requiring tenants to vote in favour of establishing a HAT to replace the local authority as landlord has had the consequence that to date only two HATs have been created.

Scottish/Welsh Enterprise Formerly known as the Scottish and Welsh Development Agencies, both of these agencies are charged with the responsibility for promoting economic development within their respective territories. Each is funded directly from the Scottish and Welsh Offices.

English Estates English Estates is a public agency that operates in economically depressed areas. It purchases land and usually develops industrial property, which it either sells or rents to occupiers. It subsidizes the process of development and therefore produces industrial units at prices below the cost of construction and often in advance of demand.

Nationalized industries Prior to the privatization process there were a number of nationalized industries that were major land-owners. Many of these have been transferred to the private sector, where they are often involved in active exploitation of their land assets, e.g. the water companies.

A number of major public land-owners still exist, e.g. British Rail, the Ministry of Defence and British Coal.

In practice, public ownership of land is quite extensive in many urban areas. Much of it was obtained by local authorities in the 1950s and 1960s by Compulsory Purchase Orders, which were used to obtain sites for housing redevelopment. Other land was obtained when local authorities acted as partners in town centre redevelopment schemes.

While public bodies have been widely criticized for under-utilizing their land holdings, much has changed since 1980. British Rail, for example, started a process of redevelopment of many of its central London stations, each of which involves the use of some of the land for office building. This commercial approach is now common, though some public bodies, most notably local authorities, may introduce non-commercial factors into their land dealings.

Private actors

Property developers Property developers are organizations that co-ordinate the processes by which a new property is created. As this usually involves risk, these organizations are invariably though not inevitably entrepreneurial. Property developers may be land-owners, builders, investors or even eventual occupiers but none of these attributes is essential. Private developers vary in size from firms with a handful of employees to major multinational organizations with assets worth billions of pounds. They seek out development opportunities, in terms of demand levels that will ensure that development will be profitable. They have to obtain land, planning permission and finance for the project and supervise the construction of the premises. Upon completion they will be responsible for letting to tenants, selling the property to an investor or, in the residential sector, selling to individual purchasers.

Investors Investors in property are those who purchase property for the returns that can be obtained from it. While the range of property investors is immense, and includes the Church, private individuals and industrial companies, the major actors are the pension funds, insurance companies, property companies and government agencies.

Pension funds and insurance companies are commonly, though somewhat misleadingly, called the financial institutions. They both obtain long-term deposits from their clients and seek profitable outlets for them. Property generally has been one of the major investment sectors (with equities and government securities), though industrial property has not been popular with them since the late 1970s. This is in part because of the poor performance of Britain's industrial sector and also because of the greater management costs associated with such property.

Property companies may invest in property as well as create it by development. They seek financial returns from property, but unlike the financial institutions they are prepared to manage their investments actively and generally incur more risk. Property companies are major investors in industrial property and are active buyers and sellers of these assets. They obtain advice from the surveying profession but also employ surveyors directly within their organizations.

The government agencies that undertake development also hold property as investors. During the 1980s they were increasingly required to obtain the maximum return possible from their investments. This is in contrast to the 1960s and 1970s when every effort was directed towards encouraging employment in the depressed industrial areas where these agencies are active. In recent years the New Town Development Corporations have been progressively abolished and their industrial property assets increasingly sold to private investors.

The privatized successors of formerly nationalized industries may also own substantial land holdings and operate in the property investment market.

In discussing the operators or entrepreneurs who operate in the property market the central rôle of the commercial property companies should now be apparent. They are not only the prime co-ordinators of the property development process but also major buyers and sellers of the existing property stock. They are the key risk-takers in a high-risk market.

Land-owners Land ownership in Britain is usually discussed in terms of freehold ownership, where the rights in land are held in perpetuity. Before the industrial revolution of the late 18th and 19th centuries, land was commonly held in very large units by a very small number of people. Some traces of this ownership pattern exist, particularly in rural areas, but much urban land has now been transferred to the ownership of industrial and financial businesses and government. The private owners are predominantly interested in obtaining the best returns possible from their land and indeed the land is considered as an investment. Profitable development is the normal investment route, but in the absence of investment opportunities the land may remain unused. Additionally, land may be held unused for considerable periods in anticipation of future development opportunities brought about perhaps by a change in the regulatory regime or perhaps by a new willingness of planning authorities to use their discretionary powers.

Financiers Financiers are those organizations that provide funding for the process of property development. Borrowing by means of the stock exchange (equity finance) has been a major source of funds for property companies. Finance has been attracted from the complete range of individuals and businesses who invest in new share issues, and as shareholders they both take the risk that the property development may fail and earn any profits if it is successful.

Borrowing by means of fixed-interest loans (debt finance) is available from the major banks and many specialized financial intermediaries. This form of borrowing is popular with property companies because it permits the profits of property development to be shared by a smaller number of shareholders.

Finance for property development has in the past been provided by major insurance companies and pension funds, but these institutions have never been the dominant actors in the industrial property market and in the later 1980s they were almost completely absent from the whole sector.

In the residential owner-occupied sector, the need for finance for house purchase has traditionally been met by the building societies. These non-profit-making bodies offer good rates of interest for small savers and lend almost solely to house buyers. For much of the post-1945 period, the building societies have been protected from competition by an advantageous tax regime, but since 1973 these benefits have been steadily reduced and competition has come from the major banks. Building societies have also been permitted to become banks and extend their range of lending. Some building societies have taken this path. They remain the main source of mortgage finance.

Users The users of commercial property, whether owners or tenants, vary from one-person businesses to major multinational companies. Their property needs are complex, changing and diverse. In Britain, commercial companies regard professional property advice as essential before long-term commitments to property occupation are undertaken.

The public In a democratic society the development process must be carried out in a manner acceptable to society as a whole. It is for this reason that development authorization is in the hands of elected authorities. There are, however, two other important ways in which public acceptability is tested. One is through the operation of pressure groups (see below), and the other is through the process known as public participation, required by the TCP Acts when new plans are prepared or certain other planning policies are developed. Public participation is not currently a major issue, though in areas where development pressures are strong the plan-making and development control systems come under greater public scrutiny.

The professions The liberal professions in the UK are incorporated in non-government institutes, usually licensed by Royal Charter to regulate qualifications for entry, professional practice and ethics, and to protect and regulate the performance of professional services, in the public interest. For each profession there is normally the requirement for entrants to have studied the subject at degree level for at least three years followed by a period of approved practice experience.

Planners The planning profession is incorporated by the Royal Town Planning Institute (RTPI), which has approximately 15,000 members, who may use the professional title "Chartered Town Planner" and the designatory letters MRTPI/FRTPI (Member/Fellow of the RTPI). The majority (around 70%) are in LPA employment, although many work as independent fee-paid professionals, often in association with surveying practices, with developers as clients. Other private sector town planners work with architecture, landscape design, civil engineering or even accountancy and management practices. Chartered town planners offer a wide variety of planning and development advice, not only in respect of the statutory TCP system but also in respect of urban and rural policy, environmental protection, economic development, tourism, etc.

The educational requirement for membership is normally at least four years' full-time undergraduate or two years' full-time postgraduate education or the part-time equivalent on a professionally accredited degree course, plus two years' approved professional practice (RTPI 1991).

Surveyors The real estate profession is incorporated by the Royal Institution of Chartered Surveyors (RICS), which has some 67,000 members. They are entitled to use the professional title "Chartered Surveyor" and the designatory letters ARICS/FRICS (Associate/Fellow of the RICS). They offer a wide range of surveying, building, estate management, property development and valuation expertise. Members may offer professional advice as independent liberal professionals (fee paid), or may be employed by property development companies or local authorities (salaried). The majority are in private employment.

The educational requirement is normally three years' higher education on a professionally recognized degree course, plus two years' approved professional practice followed by an examination known as the Test of Professional Competence.

Estate agents These may be chartered surveyors, but many firms offering a service selling or letting residential property do not employ RICS members and therefore do not offer the whole range of property services (especially in respect of the commercial sector) that would be expected of chartered surveyors. There is no legal restriction on this, only on the use of the title Chartered Surveyor. Estate agents may be members of the Incorporated Society of Valuers and Auctioneers (ISVA) and/or of the National Association of Estate Agents, both of which seek to maintain professional standards in the public interest. The ISVA sets similar educational standards and entry thresholds to those of the RICS.

In July 1991, new regulations governing the operations of any firms in the business of buying and selling houses came into force under the Estate Agents Act.

Architects The professional body serving the profession is the Royal Institute of British Architects (RIBA), which has around 30,000 members. However, to practise and retain their professional title, architects must register with the Architects Registration Council of the UK (ARCUK), a separate body set up under the Architects Registration Act. In 1993 the government announced its intention of repealing this Act, but no Bill has so far come before Parliament.

Architects designate themselves Architect RIBA or, if they are not RIBA members, Registered Architect. The rôle of the architect is that of designer of buildings and schemes for building restoration or alteration, and manager of the project through to completion. They rarely act as developers, assemble land or prepare large-scale development plans, although they may work as part of a team on such projects. The majority are in private practice but many are employed in public authorities or by commercial companies, housing developers, retailers, etc.

To qualify, normally five years' higher education on a professionally recognized course is required, followed by two years' professional practice, followed by a further examination in professional practice.

Lawyers The legal profession is split into two parts: barristers and solicitors. Barristers are a small profession, whose prime rôle is advocacy in courts of law. Until 1993 judges have always been drawn from their ranks.

The great majority of lawyers are solicitors, who are regulated by a voluntary council, the Law Society. A solicitor must hold a practising certificate issued by the Law Society and hold an indemnity insurance policy against professional negligence.

To qualify, normally a three-year law degree, plus a nine-month graduate training course and a two-year training contract with a legal practice are required. Alternatively, Common Professional Examinations set by the Law Society may be taken. In view of the differences between Scottish and English law, a practising certificate applies to one or the other and it normally takes a further training contract of one year to obtain the other certificate.

Licensed conveyancers Conveyancing of property may be carried out by licensed conveyancers (LC) instead of solicitors in England and Wales; within Scotland, solicitors retain a monopoly of property work. LCs operate under a statutory council, and are licensed following a two-year in-house training period and completion of the council's examinations. Indemnity insurance is mandatory. LCs are commonly employed by law practices but can operate independently.

Housing Housing managers have been employed in large numbers by local authority housing departments for many years without enjoying the status or having the educational level of an incorporated profession (Laffin 1986). During the 1980s, the reduced rôle of local authority housing, together with the enhanced rôle in social housing played by housing associations and private house-builders, led to a reappraisal. With government financial support, a set of one-year graduate diploma courses in Housing Management have been set up by universities and polytechnics, leading to qualifications recognized for membership of the Institute of Housing, designated MIOH.

Accountants The accountancy profession may play a rôle in urban development not just in respect of the traditional accountancy function; several firms now offer comprehensive project management services involving all financial aspects of project development and financing of development, including expertise on local, national and EC financial assistance schemes. Firms offering such services normally employ planners or surveyors in addition to staff qualified in accountancy.

Insurance, liability and ethics In some cases, professional indemnity insurance is a condition of practice and it should always be held by professionals in private practice.

All the chartered institutes have codes of practice and professional ethics to which members are required to adhere, indicating the extent of their liability and the professional standards expected, and all have power to suspend or withdraw membership from members found guilty of professional misconduct. In recent years most professions have imposed a duty to undertake continuing professional development, submitting records to ensure that practitioners keep up to date.

All chartered professions have regulations governing advertising as part of their code of conduct. In general, these specify a standard form of public notice announcing who is responsible for a project. In recent years, restrictions have been relaxed along with regulations governing demarcation of professions and the exclusive right to offer certain services. Consequently, for example, banks, accountants, solicitors, etc. may offer property for sale, often as part of a comprehensive package including mortgage, insurance or legal services.

In the UK context, no contradiction is implied by being at the same time a professional adviser and member of a chartered profession and a salaried employee of an LPA or other public body. The duty remains the same: to offer the best professional advice to their employers, whether they are private clients or local councillors.

Pressure groups A pressure group is any association or organization that seeks to influence decision-making without actually assuming the powers of government itself. A number of organizations are important in this context. It is important to note that, in Britain's discretionary system, pressure groups attempt to influence decision-making across the full range of the planning process, from plan making to development control and from central government policy guidance to very localized decisions at LPA level. The EC tier of government is another policy-making body that British pressure groups attempt to influence. Furthermore, pressure groups are generally seen as a legitimate process of public participation though it is recognized that they may not be fully representative of public opinion.

4.3 The process of implementation of plans

Implementation of development plans is not a formal process or a part of planning procedure. Development plans in the UK are statutory in the sense that they are prepared in accordance with the TCP Acts, but they are statements of policy to guide developers and residents, and inform the operation of development control by indicating the manner in which the LPA would like to see land in its area developed.

Plans are therefore informing, enabling and facilitating. Implementation may be a purely private process in the hands of the developer, in accordance with procedures and regulations outlined elsewhere.

The enabling rôle of development plans may apply to other formal procedures, for example by supporting the justification for a Compulsory Purchase Order or the submission of a development project for funding under some aspect of the Urban Programme, EC structural funds or other funding scheme where it is necessary to satisfy the fund-awarding body that the project is within the approved development plan for the area.

If a development is to be undertaken solely by the LPA itself, the LPA must still award itself planning permission, and then proceed as developer. Prior allocation of the site in a development plan, and consistency with that development plan, are not a formal legal requirement, as the discretionary principle applies to LPAs as to private applicants for planning permission, but consistency with the plan may be expected for political reasons.

The formal processes of the UK planning system are good at preventing development but not at promoting it: the system is proscriptive not prescriptive.

Planning gain

This method of positive planning is sometimes adopted to secure implementation of development as the LPA wants, especially where market conditions and current plans give the LPA a strong negotiating position. An agreement is made under section 106 of the TCP Act 1990 (sometimes still referred to as a "section 52 agreements" after the relevant section of the 1971 Act). This is an agreement whereby, in association with the granting of planning permission, the applicant may agree to the phasing of development in accordance with the anticipated provision of infrastructure in order to avoid extra burdens on the local authority, or to provide some space laid out as public open space for recreation, or make a financial contribution to the cost of providing roads, drainage, a school or other infrastructure so that the development can proceed without delay.

Agreements may be with other public authorities as well as the LPA under section 106 of the TCP Act. They bind land-owners and their successors in title and are registered on the legal title of the land affected.

Planning gain, or developer agreements, were increasingly widespread at the end of the 1980s, especially in parts of the country where property values and potential profits from development were high, because the ability of local authorities to finance the necessary social and physical infrastructure was very limited owing to cuts in public expenditure allocations. The extent to which LPAs can or should legitimately negotiate planning gain in the form of material benefit to their community is somewhat controversial. Some LPAs tried to exploit their bargaining power to maximize the developer contributions to their expenditure, though most used the powers to restrict development (Ennis et al. 1993).

Section 106 agreements, or the likelihood of their being sought by the LPA, are reflected in the land market and prices, and add to the uncertainty of any developer or purchaser prior to obtaining planning permission.

The Planning and Compensation Act (1991) replaces section 106 in the 1990 Act with a new section. It replaces the term "planning gain" with the term "planning obligation" and spells out in more precise detail what is possible and in what circumstances (Grimley J. R. Eve 1992).

CHAPTER 5
Ownership

5.1 Changes in the structure of ownership

Land

Very little is known about changes in the ownership of land. The absence of a cadastre ensures that research is restricted to specific locations and that generalization would be speculative.

One general assumption is that, as far as potentially developable land is concerned, there will be few land-owners who are not aware of the increase in land value that would occur if planning permission were available.

Housing

The issue of housing ownership has been discussed in Chapter 3.3 and elsewhere. While the growth of home ownership is one of the principal characteristics of Britain's housing market, it is important to recognize that cultural and other factors do influence ownership patterns. For example, in Scotland a tradition of tenanted dwellings has persisted and where the private sector could not meet local needs the public sector has provided an extensive housing sector. The Scots have been far less willing to participate in the right-to-buy scheme than their English counterparts.

Another example exists in rural areas close to urban centres where the availability of housing in a pleasant environment has resulted in whole villages being converted to high-quality owner-occupation at the expense of low-paid agricultural workers, who often face real difficulties finding accommodation close to their employment.

The general point is that housing tenure is influenced by a range of factors, and space is not available here to give a full explanation of the various patterns.

Lastly it should be noted that the issue of inherited wealth from house ownership is now becoming a major financial issue and an increasingly widespread source of capital for second-generation home-owners. It is leading to the purchase of second homes in attractive rural areas such as Wales and the Lake District and overseas.

Commercial Property

Very little is known about the pattern of ownership of commercial property. Market information on property purchases and sales provides evidence on changes in ownership, but the underlying pattern is difficult to discern, especially for secondary property. There is little doubt, however, that the dominance of the investment market over the owner-occupied market has ensured that every property-owner is aware of the investment value of their property interest. Property is treated as an asset, to be bought, sold or developed, rather than as a means of production. Many large industrial firms are well aware that more profits can be made from careful exploitation of their property assets than from their main business. Not surprisingly they often appear to concentrate more on property management than on producing goods and services.

5.2 Demand for and supply of building land

Local authorities maintain registers of land that has received planning permission for development in each year. These are not collated centrally, though some estimates do exist. Detailed research is done in the private sector, where it is seen as valuable information to market actors.

For example, in the London office market (see Ch.9.1) information is available on the developments that have begun and on the sites that have received planning permission. The market professional can therefore predict quite accurately the new supply on the market as current building activity reaches fruition and they can predict potential future supply. As a by-product, information on building land is available. Outside London the information is far less valuable and therefore often not collected.

Where the public sector is involved in land assembly more information is available. This is particularly true of Enterprise Zones and Urban Development Corporations. While the areas covered by these schemes are quite small, the use of public subsidies has ensured that, outside London, they have attracted much local development and they are therefore more important than they at first appear. For example, in Newcastle there has been an almost complete absence of commercial development activity outside of these areas.

The issue of housing land has been contentious in Britain and a housing land availability mechanism was created (see Ch.3.1). Quite accurate local measures of land being supplied for housing were therefore available and the process was intended to ensure that housing land supply matches housing demand. The new Unitary Development Plans and district-wide local plans should ensure that the supply of land for housing is maintained.

The demand for land, and its price, in Britain are tied to the property market, where cyclical patterns exist. The figures for money into property (see Table 1.15) illustrate the point. Rising property prices lead to increases in land prices at rates over and above the increase in property prices because much of the increasing profitability of development is translated to the land market. When property prices fall, land values fall dramatically and, when the value of a property reaches its development cost, then in theory the value of land is zero, though in practice transactions simply cease.

Information on the numbers of transactions in the land market does not exist, though professionals who work in these markets are well aware of the speed of local transactions. They are also aware of local prices and land availability for each sector of the market.

5.3 Prices

The information on land and property prices is included in Chapter 1.4 There are clear differences in price levels, both sectorally and spatially.

Sectoral differences

A sectoral hierarchy exists in that the highest rents are paid for retail sites in prime locations, followed by offices, industrial premises and housing. Thus generally retailing is able to outbid all other land-uses for available land. The general pattern is, however, subject to many anomalies. For example, in some good-quality urban locations housing is able to outbid industry, while in other areas retailing would be unprofitable and hence it does not enter into the calculations.

Actual rent levels and land usage are conditioned by the existence of the planning system and, while it is in principle a discretionary system, there is little doubt that the system does restrict some forms of development in some locations. Land values are therefore strongly influenced by expected planning permissions. For example, the strong policy stance taken in supporting the maintenance of the green belt around major conurbations ensures that land prices do not vary far from agricultural prices except in special cases.

Spatial Differences

Spatial differences in land and property values are widespread and substantial. Data are included in Chapter 1.4.

The key concept is whether the land can be developed profitably. If the development value determined in the user market (rents) and investment market (yields) exceeds the cost of the development, then the private sector will, subject to planning permission, create the development. This deterministic explanation must be modified by the recognition that property developers face uncertainty as to the future levels of property values and building costs and consequently actual decisions depend upon expectations.

Where value exceeds the costs of construction, the land value will be positive and it will generally reflect the profitability of the land-use (see Ch.4.1).

Transactions

The level of prices and the number of transactions are clearly connected in Britain's housing market, though difficulties in comprehension are created by the persistence of inflation in Britain. The data in Table 5.1 indicate that a house price boom peaked in 1980–1 but the level of transactions peaked in 1978.

Table 5.1 The Transaction Model, 1975–92.

Year	Building Society advances ('000)	Dwelling prices £
1975	652	12,057
1976	717	12,906
1977	737	13,764
1978	802	16,026
1979	715	21047
1980	675	24,307
1981	736	24,810
1982	861	25,553
1983	950	28,592
1984	1,086	30,811
1985	1,073	33,187
1986	1,231	38,121
1987	1,051	44,220
1988	1,231	54,280
1989	871	62,135
1990	775	57,869
1991	721	56,889
1992	701	52,829

Source: Housing and Construction Statistics and Nationwide Building Society.

From 1981 the level of transactions expanded steadily until 1986. During 1987 the level of transactions started to fall, but they rose again in 1988. This exception to the transaction model in Table 5.1 could be explained by a sudden increase in sales associated with the imminent ending of certain tax advantages.

From 1988 onwards the level of transactions fell rapidly but price inflation continued into 1989. House prices fell during the early 1990s but had stabilized by 1993. Transactions in 1993 were starting to rise.

5.4 Speculation in land and property

The common notion of speculation
The concept of speculation in land and property is difficult to define in Britain. At certain times, particularly when land and/or property prices are rising very rapidly, there has been some public disquiet directed at those who have "earned" large sums by "sleeping while their assets rise in value" and "who have made no real contribution to society". This attitude sits uneasily with the reality that a large proportion of the population are owner-occupiers and have made similar gains with the increasing value of their own homes.

Developers and investors in property are seen as entrepreneurs who take risks in the hope of realizing profits, as people in any other business do. Speculation in this sense is normal and accepted; it is only if there is any suggestion of market manipulation, inside knowledge or excessive windfall gains that it is in any way unacceptable.

More generally, the public view is that the decision to purchase land or property should be taken in the light of possible future changes in value; it should be seen as a rational and acceptable decision. At the personal level, it is normal for house purchasers to perceive their property as a "good investment" – an expression that can be interpreted as a property whose value should rise more than the average.

In the property investment market it is possible to argue that all invest-ments are to some extent speculation, but the term speculation is not used. Rather, the investment of cash in property is seen as a respectable activity and as a normal way of diversifying the risks inherent in other investment sectors.

Commercial firms that purchase land and property as part of the development process are perceived as more questionable, but successful investments are generally seen as shrewd business decisions.

The only form of land/property purchase that is really seen as unaccept-able speculation is land banking (discussed in Ch.3.2). Even that process is recognized by many as inevitable in a market economy, and it is only the most speculative of purchases undertaken purely in the hope of obtaining planning permission in Britain's discretionary system that are regarded with some disquiet.

The extent of speculation

Speculation in housing exists to the extent that purchasers anticipate future changes in property values and national prices and incomes. It is not uncommon for people to borrow more than they can really afford to repay in the expectation that rising money incomes will reduce the debt in future years. In markets where house prices are rising rapidly, some purchasers buy quickly before prices rise further and by doing so increase the rate of house price inflation.

From early 1990 until 1993 the level of house prices fell quickly, with some regions recording falls of 40%. This has left many householders with negative equity and they recognize that they have made a poor investment decision. In their defence it must be said that this is the first time that house prices have fallen on this scale in the post-1945 era.

If the land banking process is seen as speculation then it dominates the land assembly process, for almost all developable land is purchased well before it is required for development. It is not seen as speculation in Britain and nor is the "normal" process of investment in land and property.

Who speculates

In housing it is clearly the owner-occupier who is the primary speculator, though in the privately rented sector some investment in property is seen to be based on speculation over the property's future value rather than on the expected rental income.

In the commercial sector it is the developers and investors who speculate. The developers are a diverse group (see Ch.4.2). House-builders are active land bankers in possible residential areas, while many types of developer work elsewhere. This activity is seen as investment rather than speculation and generally does not engender hostile comments.

Risks

In the housing sector, the clear and obvious risk is that property is purchased at the top of the price boom and a capital loss results if the property is sold at a later date. Given that building societies and banks have been willing to lend up to 100% of the house value to some purchasers, the issue of negative equity and housing repossessions was important in 1991-3. The risks of large-scale default could become severe.

Another form of risk occurs in the owner-occupied sector. When the market is depressed it may be very difficult to move house in search of work. Furthermore, it may be very difficult generally to move out of a depressed region because house prices in other regions tend to be higher. Today this physical immobility is seen as being a major constraint on labour mobility and a risk that home ownership entails.

In the property development sphere, the purchase of land at too high a price carries the risk that the development would not be viable. Developers do make such mistakes, and when there is a downturn in the economy a large number of bankruptcies amongst developers do occur. At certain times (1973/4 and 1992/3) the scale of failure of property loans can have a major impact on the financial system. For developers, the risks associated with the property cycle are seen as part of the process and of course each firm attempts to avoid them by thoughtful purchasing.

In the investment market, the rise and fall of property values are seen as part of the investment process. Many investors hold property for very long periods of time and are therefore not greatly concerned with short-term variations in value. Indeed, investment into property is really not seen as speculation; rather it is the management of the risk/return trade-off. For those investors who buy property at the peak of the property cycle the losses can be severe and damaging.

CHAPTER 6

Case studies of the land market

6.1 Stockley Park: Business Park, London Borough of Hillingdon

This Business Park has been developed since the mid-1980s and is perhaps the best British example of up-market high-quality industrial premises. It is situated 2 miles north of Heathrow Airport in west London and adjacent to the M4 and M25 motorways The London Borough of Hillingdon is the LPA.

Prior to the development, the land was owned in part by a private company (80%) and in part by the local authority (20%), which had used part of the site for refuse disposal. The land was seriously derelict and the need for reclamation became clear after two fires in the rubbish had been extinguished at considerable expense.

Reclamation by the local authority would have been expensive because this was a large site and central government funds were not available. Hillingdon Council had previous experience of this type of problem on another site and in that case they had used the profits from eventual housing development to fund the financing of the land improvement.

Hillingdon decided to approach this site in the same manner and its initial strategy envisaged the creation of warehouses on part of the site, the profits from which would fund the site improvements. This proposal was made despite the land being part of the green belt, which gave the land the highest form of protection against development activity within the planning system.

Hillingdon established 11 principles in 1980–2 to guide the development of the site. These were:
– release as little land as possible to generate funds to restore the site;
– ensure that land release did not make green belt boundaries less defensible;
– restore the site so that it was safe for public access and could support green belt compatible land-uses;
– ensure public access to the site;
– improve visual appearance and amenity of the site;

- widen the range of employment opportunities in Hillingdon;
- realize the recreational potential of the adjacent canal;
- relocate existing non-conforming users;
  improve transport infrastructure for the benefit of the area as a whole;
- stop the site being a source of pollution;
- make the site safe.

At about this time the adjacent private land was sold to its present owners, a property development company. From its point of view the site was almost ideal, being very close to London Heathrow Airport and major motorways. It had obvious attractions to international companies and, if a high-quality development could be created it would be attractive to managerial and highly skilled employees. The developer managed to persuade the local planning authority that it would be in the best interests of the local community if a larger, higher-quality development were permitted on a different part of the site. Such a development was possible because of the rising level of demand for such developments in the locality.

Hillingdon responded by accepting the developer's strategy, but it wished to ensure that there would be public benefits from this development (planning gain) and that the development of the site would be restricted to industrial development. Eventually it was agreed that both the developer's and Hillingdon's requirements could be met by a section 52 agreement under the 1971 Act (now known as a section 106 agreement under the 1990 Act).

The proposal

The proposal agreed by the developer and the local authority in both its rôles as land-owner and planning authority was for an industrial development involving:
- 45,500 m² industrial space;
- 45,500 m² warehousing;
- 45,500 m² ancillary office space.

The local authority wished to ensure that the development did not take the form of an office park and consequently vetted the proposed tenants to ensure that the office use was truly ancillary to the other uses on the site. In addition, 25% of the Business Park area is a green landscaped area.

The developer agreed to an extensive programme of site improvements to the remainder of the site. The agreement covered the method of restoration and the resultant land-uses: golf course, woodland, parkland, recreational pathways, etc. When the work was complete and the site fully available for public use, the land was transferred into public ownership.

The final form of development is quite unlike that originally proposed because during the initial stages the government changed the development regulations by changing the Use Classes Order that permitted the creation of B1 developments (see appendix). The local authority was forced to accept that industrial land could be used for office activities and in effect had to accept a far higher level of office use than it had previously desired.

The Business Park

The Business Park covers approximately 40 ha of the 140 ha site. The developer has provided shell and core buildings, leaving the occupiers to fit out the premises to their individual specifications. The buildings are provided on a leasehold basis, are of two or three storeys, have generous ceiling heights for air-conditioning systems and are provided with ample car parking facilities. One particular feature of the site is the extensive use of water as a means of creating an attractive environment.

Over 60% of Stockley Park is occupied by non-British firms and it commonly accommodates company head offices, research and development centres and computerized production facilities. The emphasis is on skilled staff, technology and decision-making.

To ensure that the site is maintained at the highest possible quality, the developer provides an active site management team. In addition to maintaining security and environmental quality, the management team oversees the provision of sports, retail and banking facilities for the employees located in the Park.

Conclusion

Overall Stockley Park illustrates a number of important points.

First, if a development proposal is judged by the LPA to be in the best interests of the community then it is possible that planning permission can be obtained even where the development implies the overturning of one of the most "sacred" planning policies.

Such decisions may be made at local level but in such circumstances central government will still have to give its tacit approval. If the local authority were to refuse the application, then the Secretary of State for the Environment may overrule any local objection after receiving the results of a planning inquiry. The decision should be made in the "public interest", though in this case the preservation of the green belt is also a "public interest" matter.

Secondly, the case study illustrates the possibility of obtaining permission for a development that is not included within the local plan. In this case the local authority actually made the first proposal to overrule its own plan, which is by no means unusual although it is more common for developers to make such an application. In either case it is the responsibility of the local authority to judge the application in the light of the benefits and costs to the local community.

Thirdly, the case study provides an example of the capture of community benefit (planning gain) by means of a legal agreement between the planning authority and the developer. Whether the planning authority could have obtained further benefits is always debatable, as is the general question of the extent to which undesirable applications are approved by the "bribery" of social benefits. Additional problems occur where the local authority is also a land-owner and would receive additional financial benefits if the development application were approved.

Fourthly, the development shows that it is possible to improve derelict land in an urban area by property development. It must be noted that it was the highly profitable nature of the development in this prime high-value location that enabled the developer to fund the land-improvement process.

Fifthly, the case study illustrates the way in which the planning system is able to respond to market trends. The ability to override the local plan enables the planning authority to adjust to changes in demand for land without the process of creating a new plan. The adjustment in the Use Classes Order has enabled developers to adjust land-use further after development permission has been given.

Lastly, Stockley Park illustrates (in an extreme form) the type of development that became increasingly common during the 1980s in Britain. Although traditional industrial units were still constructed, especially in older industrial areas, the growth sector of the market has been better-quality premises with a high proportion of office space. This change has been facilitated by the creation of the B1 Use Class but more importantly it reflects the changing nature of industrial property demand.

6.2 Newcastle Business Park: Tyne and Wear Urban Development Corporation

This case study examines a major office and light commercial development by the private sector in partnership with the public sector through an Urban Development Corporation. The development occurred on derelict land within the local government district of Newcastle upon Tyne and would not have proceeded in its present form without public assistance to bridge the gap between development cost and developed value.

The principal parties in the development of the Business Park are the Tyne and Wear Urban Development Corporation (TWDC) and Dysart Developments (Tyne and Wear) Limited (Dysart). The TWDC was established in 1987 with resources of £200M for its first six years. Its designated area covers some 2450 ha, predominantly along 43 km of the River Tyne flowing between Newcastle and Gateshead and the River Wear flowing through Sunderland 19 km to the southeast, of which over a third was derelict land. The rôle of the UDC is explained in Chapter 3.3. Dysart is a small development company controlled by a husband and wife partnership. Until the Business Park development, it was primarily known for retail development in the south of England. Dysart's funding arrangements are confidential but sourced from one lender.

Industrial history of Newcastle

Newcastle was at the forefront of the industrial revolution and in the 19th century was a major centre of coal production, ship building and heavy engineering activities. The decline in these industries through to the 1980s severely affected the local economy, throwing it into a depression and leaving a legacy of dereliction and pollution (Robinson 1988). In 1993, closure of the last major shipbuilding yard, Swan Hunters in Wallsend, is likely to occur. Today, Newcastle is primarily an administrative and educational centre, with the public sector being the major employer, and has a much reduced scale of industrial activity.

Demographic information

Newcastle is generally recognized as the regional capital of the northeast of England and the local authority district of Newcastle has a population of 279,800. The total population of the former metropolitan county of Tyne and Wear, which contained five districts, is approximately 1.25M.

Other commercial developments and rents

Office accommodation in the northeast of England is concentrated in Newcastle. By 1985 there was approximately 372,000 m² of available office space in the city, 75% of it located within the city centre and the remainder being mainly located just to the north of the city centre in the areas of Jesmond and Gosforth. That Newcastle should remain the centre of office space was a policy of the Tyne and Wear structure plan, with Sunderland being seen as a secondary office centre of the region.

Office development and rents were static in the period 1980-6, the rental level being around £43/m². The economic boom of the southeast of England towards the early and mid-1980s spread to the region by mid-1986. Demand for office space soon exhausted available supply and rent levels doubled to range from £81 to £110/m² by the end of 1989. Prime office space in the centre of Newcastle reached £129-151/m². Development activity intensified at this time and 325,000 m² additional office space was proposed mainly by the private sector from 1988 to the early 1990s.

Background to the Business Park project

Prior to the formation of the TWDC in 1987, the freehold to the site and the land immediately to the east was held by the City of Newcastle District Council. In seeking to promote development of the site, the council sought its inclusion within an Enterprise Zone (EZ). This was achieved on 25 August 1981, with planning permission for Class B1 development.

The city council was keen to develop the site as a counterweight to the MetroCentre being planned on the opposite side of the Tyne, also in the EZ. In 1984 the city council sought development proposals from the private sector and, of the 32 developers participating, selected a proposal submitted by Dysart Developments (Tyne & Wear) Limited. The selected proposal was for a mixed development of industrial park, business park, riverside village, retail units and leisure facilities to be completed in partnership with the council and subsidized from public funds. The scale of the site and the absence of massive public sector subsidy at that time dictated a piecemeal approach to the development, with decontamination of the soil and provision of infrastructure being completed in stages as building progressed and occupiers were found. By mid-1986, subsidized by Derelict Land Grant worth £2M, available then only to the public sector, industrial units to the east of the site were completed and occupied. Work commenced on this part of the project first because the existing access roads were already adequate for that use.

At the heart of the financing of the whole development was the revenue to be generated from the retail units. Unfortunately, in 1986 that particular sector slumped and the completion of the entire scheme was undermined as financing centred on anticipated income from the retail units.

This event coincided with the formation of the TWDC in 1987. The developer's and the city council's initial reaction was to proceed as planned but to seek assistance from the TWDC for the provision of the infrastructure. This was not acceptable to the DoE. Discussions resulted in the developer withdrawing from its partnership with the city council and the TWDC acquiring the site for development under new proposals to be provided by the developer. The new proposals were required because the retail and leisure elements were considered inappropriate by the TWDC. The developer returned to the TWDC with proposals for a business park for office and light industrial use. These were acceptable and planning permission given by the TWDC as the LPA.

The Business Park scheme

A formal agreement for development was completed between TWDC and Dysart in July 1989. This committed Dysart to a development timetable that in effect made it a speculative development. Ownership of the land would remain with the TWDC until the lease or sale of the units to occupiers, with Dysart as consenters for value. Tensions were mentioned by Dysart in the ensuing year of the agreement; Dysart was concerned with the commercial sense of securing commitment of a tenant prior to construction to meet desired specifications and TWDC wished for completed development for demonstrable progress in its publicized activities. Strong market demand

Table 6.1 Newcastle Business Park: timetable of events

1970s	Ownership of site acquired by District Council
1981	Inclusion of site within Enterprise Zone.
Early 1985	Development partnership established between District Council and Dysart
Mid 1986	Industrial development completed on adjacent site under that agreement
Early 1987	Partnership faltered in retail slump
April 1987	Formation of TWDC
Late 1987	Development halted and discussions undertaken with TWDC
July 1989	District Council withdrew and formal development completed between TWDC and Dysart, with title of site passing to TWDC
Early 1991	Completion of development activities
Late 1991	Anticipated sale of residual interests in development to institutional investor

prevented any breaching of the agreement and the majority of the units achieved occupiers prior to construction. A breach of the agreement on the part of Dysart would have enabled the TWDC to put it off site and proceed with an alternative developer.

TWDC's rôle was in the preparation of the site, extensive landscaping, provision of infrastructure and greening of the site through planting and provision of street furniture. Assistance worth £12.5M was provided by the TWDC, total project costs being £141M.

The site of the Newcastle Business Park extends to 24 ha and is located 2 km to the west of the city centre. It is bounded on the south for 1.5 km by the River Tyne, on the north by a major arterial road, on the east by an estate of small industrial units and on the west by established heavy industry. The site has been landscaped extensively onto two levels owing to its steep incline towards the river. Just over 37,160 m^2 of high-quality office and light manufacturing space has been constructed at a low density within a landscaped parkland. A variety of units have been built, ranging from 24 terraced office suites of 112 m^2 each, 6 at 168 m^2 and 16 larger units ranging from 465 m^2 to 10,680 m^2. This development is supplemented by a range of small service units providing shopping and eating facilities for site users.

Prior to development, the site was derelict, having been the site of Armstrong's Elswick works whose heavy industrial activities employed 25,000 people at the turn of the century. Cessation of these activities in the 1970s left disused buildings and an unstable and polluted soil. The transformation and improvement of the environment are dramatic and offer an extremely attractive and modern working location.

Marketing of the units was a function shared by TWDC and Dysart. Dysart's expenditure on marketing was £690,000. Direct lobbying of potential tenants and promotional events were supplemented by advertising in the professional press and exposure through firms of chartered surveyors. Marketing by TWDC tended to be part of general promotion activities of its work in its designated area. Uptake of the units was swift, with $36,500m^2$ being let in a period of just over 12 months.

Dysart's rôle in the Business Park will cease on the disposal of the final unit. The landlord interest in the leased units have been to be sold to an institutional investor and a management company has been established for maintenance of the general amenities of the Park.

The outcome

The Business Park offers high-quality office accommodation in a uniquely attractive environment. Occupiers were attracted by public sector subsidies. The most important of these, until August 1991, was its EZ status, with exemption from the Uniform Business Rate tax and 100% capital allowances against Income and Corporation Taxes. This was supplemented by grant and loan assistance available through TWDC, mainly in the form of assistance towards relocating, and training and arranging financial packages between the public and private sectors. The area of Newcastle also enjoys development status, attracting central government assistance in the form of City Grant (see Ch.3.3) and ERDF.

All units on the Business Park were built to offer flexible accommodation for future requirements of occupiers and raised flooring for communication systems. State of the art telecommunications were installed through the Park by British Telecom.

Rents during the main letting period of 1989 ranged from $£102/m^2$ to $£110/m^2$, including a service charge of $£8/m^2$. Units were leased on a full repairing and insuring basis. A number of units were purchased.

The collapse of the property boom affected Newcastle in early 1990 and further office developments are being rescheduled. All but the small terraced office space is let or sold on the Business Park and it was opportune that this was achieved prior to the onset of the recession.

Many prestigious companies now occupy the site, including British Airways' central reservation facilities, the Agricultural Intervention Board (a central government agency), IBM, AA Insurance Services (a major motor insurer), Cellnet, and Merz and McLellan (Engineers).

Summary

1. Over 24 ha of derelict land has been transformed into office accommodation suitable for high-technology users into the 21st century.

2. An initial public sector investment of £12.5M generated a private investment of £128.5M, which would not have proceeded otherwise.
3. The development will create 5000 jobs when it is completed, in an area of high unemployment. This project demonstrates how one of the public policy instruments of the 1980s (UDCs) has enabled the development of urban derelict land.
4. It worked in partnership with the private sector to improve and develop the site. Initially the UDC was prepared to accept that it should put £10M into a project worth only £40M, a leverage ratio of 1:3.
5. The project was much more successful than originally anticipated and the developer eventually invested much more into the development.
6. This was made possible by the very flexible planning regime created within the UDC.
7. It was facilitated by the buoyant property market during the development stage.
8. The TWDC was concerned to ensure that the development proceeded as quickly as possible because it wished to demonstrate a "success" attributable to itself. In practice this did not obstruct the process of development. The developer was very prepared to work with TWDC and recognized at the outset that development would not be possible without public subsidy.
9. Dysart saw its relationship with TWDC as one of bringing respectability to the developer as well as creating planning and financial benefits.

6.3 Aberdeen: greenfield housing development on Lower Deeside

This is a study of a residential development of 40 units by the private sector on a greenfield site. The development site is on the periphery of the commuter settlement of Bieldside, 9–10 km west of the city of Aberdeen in the City of Aberdeen District of Grampian Region in Scotland.

Grampian Region has a population of just over 500,000. Bounded to the north and east by the North Sea and with an abundance of good agricultural land, fishing, agriculture and whisky production have been the traditional foundations of the region's economy, with modest industrial activity mainly centred in Aberdeen. Aberdeen is by far the largest settlement on the east coast and is the economic, administrative and cultural centre of the region. With Peterhead to the north, Aberdeen is the biggest centre of whitefishery in Europe. As with other large Scottish cities, the majority of the older residential accommodation in the city is in the form of flats, bringing down the average dwelling price in the city.

Grampian would have remained a typically peripheral rural area but for the discovery of oil and gas in the North Sea in 1969. Existing harbour, airport, rail, road and education facilities, coupled with a responsive and sympathetic local government, meant that Aberdeen developed through the 1970s and 1980s into a centre for oil and gas offshore exploration and production activities, and associated administration. Although always enjoying modest prosperity, the economy of the region was greatly enhanced by this development which was reflected in 1991 in a regional unemployment figure of 3.4%. Direct regional employment in the oil and gas industry is currently 52,500, or 20% of total employment. However, 26,200 of these jobs are on offshore installations with only one-third of the workers residing in the area. It is thought that a further 20,000 jobs are directly dependent on the oil and gas industry. The service sector, employing 120,000 or approximately 45% of the workforce, remains the largest employment sector.

Bieldside is one of a string of predominantly commuter settlements to the west of Aberdeen along the north side of the Dee Valley, the area generally being known as Lower Deeside. An area of natural beauty, away from the windswept east coast, it has for a considerable number of years been a prime residential area. There is a marked absence of local industry or shopping facilities for other than local needs, and the housing is of the highest quality in the Aberdeen area. The study site, which extends to 3.36 ha, lies on the north of the settlement and extends northwards to the line of the green belt. It is flanked on the west by recently completed residential development of similar-quality housing and on the east by local education facilities. The south-facing site slopes gently, and at an elevation of some 80 m above sea level has commanding views southwards across the Dee Valley.

In 1975, local government was reorganized, with the formation of Grampian Regional Council subdivided into five district councils. Strategic planning is undertaken by the region and the first structure plan covering Aberdeen and a radius of 30 km was published in 1981 and reviewed in 1986; a further review was initiated in 1991. Policies and proposals contained in the structure plan on the basis of its survey findings, including housing allocations within established settlements, are implemented and given substance in the local plans prepared by the district councils, which also make the individual development control decisions against the framework of the structure and local plan.

The Scottish planning system differs from that of the rest of the UK in that it has for several years required complete coverage of local plans. The first local plan for the study site was published by Aberdeen District Council in December 1982 and, following a Public Local Inquiry, was finalized and adopted in September 1984. A review is in the course of being finalized.

In accordance with government policy expressed in Circulars and Policy

Guidance notes issued by the Scottish Office, the regional council is obliged in preparing its structure plan to ensure that there is an effective five-year supply of building land, based on its survey of the area and its projected growth. The effective supply was established by reference to the remaining capacity on sites under construction, sites with planning permission and allocations made under existing local plans, with a discount being made for development constraints existing over that five-year period. In order to ensure the five-year supply, close liaison with the building industry is maintained. Consultations between Grampian Region and a body known as the Builders' Forum, representing the house-builders operating in Grampian, take place at frequent and regular intervals. A detailed analysis of housing and also industrial developments by site and settlement is prepared annually by the region in connection with this function.

It is important to note, however, that allocations made in the structure plan do not always accord with locations sought by builders. This is best illustrated in the case of Cove Bay, a village immediately to the south of Aberdeen on the coast and perceived by the district council as a suitable site for natural expansion of the city. Land is allocated there in the local plan for over 1000 houses. That allocation has not, however, been taken up, the allocated sites being regarded as too exposed to the wind and rain from the sea. Pressure for development on inland sites, particularly in Lower Deeside, therefore continues.

The 1986 structure plan identified an effective supply of 450 houses for the Lower Deeside area and made an allowance for 250 additional houses to be allocated in the period to 1996 in the local plan.

Development history of site

In the early 1980s development planning had fallen behind economic events in the region. Demand for housing, particularly in Lower Deeside, was outstripping supply. Cala Homes, a national builder, made an application in 1981 to develop 20.9 ha to the west of the study site, but the district council refused the application on the basis that the local plan at that time was at a critical phase in its preparation and that it would be inappropriate to approve the application ahead of public consultation in the plan-making process. An appeal was therefore made to the SOS for Scotland. The Reporter (Scottish equivalent of Inspector) approved part of the application but indicated that approval of the remainder of the appeal site would be given in the future, the infill of land up to the green belt boundary being a desirable development and application of the housing allocation then provided for in the structure plan.

This forced the hand of the LPA: in the adopted local plan the appeal site was subsequently designated for housing development. The decision raised optimism in Cala that the study site would also in time receive planning permission, being infill of low-grade agricultural land to the line of the green belt. The site was acquired speculatively and allowed to lie fallow as part of its land bank.

Following the planning appeal decision and anticipating residential planning applications for the remainder of the appeal site, the district council prepared a design brief for the guidance of developers. The brief was issued in October 1984 and extended beyond the appeal site to the study site. The brief advised on housing density and the requirements that would need to be met for provision of open space and infrastructure. This signal of a positive attitude to a planning application was confirmed as the district prepared and issued its draft local plan wherein the study site was provisionally designated for residential development.

A slump in the housing market in 1986 (discussed below) resulted in Cala ceasing operations in the Aberdeen area and the site was acquired by Headlands Homes in late 1988 at a price of £587,000. This amounted to £174,702 per ha, or approximately £15,000 per dwelling. In acquiring the site Headland was confident of obtaining planning permission at a density of 12 units per ha, that confidence stemming from the contents of the design brief and good communications established by it with the District Council, which was initiating a review of its local plan. Headland's confidence was justified, and expressed in the absence of any conditions to the sale agreement requiring planning permission to be granted prior to completion of the sale, an otherwise common contractual condition.

An application for full planning permission was submitted by Headland on 4 April 1989 and received approval on 8 August 1989. Planning applications are normally made in outline but the clear statements of the district council through the design brief, and close liaison between developer and LPA during pre-planning application negotiations, enabled Headland to jump this stage. This, however, is an unusual occurrence.

Having obtained building warrants and complied with requirements for the deposit of a bond against completion of the roadways to local authority standards, Headland started construction work in September 1989 as the house market recovered from its slump. It had correctly anticipated from its market research that there was a suppressed demand for quality four-bedroom accommodation.

Constructing in two phases of 12 and 28 dwellings, Headland achieved sales in almost all cases ahead of completion. The first house sold was reserved in July 1989, prior to the start on site, and was occupied in March 1990. Sales were generated through the press, a sales office situated on the main street of Aberdeen and on the site.

Prospective purchasers, in accordance with normal builders' practice, made a small deposit of £100 on intimating an interest, with exchange of "missives" (the Scottish terminology for the exchange of a binding contract to purchase) being concluded within 2-3 weeks. Missives were in some cases concluded ahead of work commencing on the house in view of the strong demand. Most purchasers were buying with a view to owner-occupation, but there was also some interest from oil companies purchasing in order to be able to offer their employees suitable short-stay accommodation in the area, and from private buyers purchasing property as an investment, intending to let to tenants.

Forty four-bedroom detached housing units have been constructed of single, one-and-a-half and two storeys. All have double garages and are of a superior quality aimed at the upper end of the housing market. Prices on the development ranged from £120,000 to £186,000 dependent on house type and plot size. Prices were fixed by Headland and were non-negotiable. A purchaser would, however, require an independent valuation of the house for mortgage purposes. The missives are a standard offer prepared by the builder and accepted sometimes after negotiated modification by the purchaser, normally but not necessarily through their solicitor.

Legal environment

Scotland has a feudal system of land tenure, all land in theory deriving from the Crown and held in a descending series of contractual grants for an indefinite period but capable of termination on breach of the contractual terms, yet freely assignable by the grantee. Historically grants were for performance of a service but by the 16th century had developed into grants for payment of an annual consideration, known as a feu duty. Feu duties were abolished in 1974 but the tenure remains and is adept in allowing the imposition of positive obligations on land, unlike the English land system where only negative obligations may be imposed by covenants.

The title deeds for purchasers on the site all contained identical conditions designed to protect the general amenity of the site, for example by prohibiting business use of dwellings and extension without the consent of the "superior" in title. Headland, as granter of the "feu", is in the position of the "superior", but the conditions may be enforced by any occupier of the site against another occupier.

House prices

The introduction of the oil and gas industry to the region and its subsequent fortunes have dominated the demand for housing in the region and the level of house prices. The oil price slumped in 1986 and the industry slipped into a severe recession, pushing unemployment up to 11.2% with many workers leaving the region. Until that time typically high-earning workers had migrated to the region, swelling its population by some 14%. A strong local demand for housing also occurred during that period and massive house-building activity occurred that increased the total stock within the region by 30%. In the late 1970s and early 1980s, house prices in the Aberdeen area were rising at an average rate of 18% per annum and were the highest in the UK after London and the South East.

The recession had a dramatic effect on the housing market and the construction industry, private house-builders being geared at that time to record levels of output, Employment in the oil and gas industry has now returned to pre-1986 levels, but only very modest net migration into the region is being anticipated. Confidence has returned to the market but with a new caution absent in the "boom" years. Table 6.2 shows that Aberdeen prices were above the UK average in the mid-1980s boom, but have since moved closer to the Scottish average. These figures are well below Lower Deeside prices, of course, as the average reflects the large proportion of apartments in the central parts of the City of Aberdeen.

Table 6.2 Average selling prices of housing, 1984–90 (£).

	Aberdeen area	Scotland	UK
1984	39,325	29,125	30,950
1985	41,025	30,000	32,950
1986	39,075	31,050	38,700
1987	38,750	35,900	47,500
1988	41,250	40,200	57,600
1989	46,850	46,000	61,150
1990	54,275	53,950	64,725

Source: Aberdeen Solicitors Property Centre, 1991.

Developers and builders

In the 14-year period from 1975, 61% of all private sector house building was undertaken by 12 private companies, but 2 companies dominated the scene, contributing over 36% of all building, these being the national builders Barratt (22%) and Wimpey (14%). The developer of the study site is Headland Properties Ltd, a locally based company with £20M sales receipts in 1990. It is a member of the Stewart Milne Group Ltd and the timber frames for the units were provided by its sister company Wellgrove Timber Systems Ltd, with construction being undertaken by its other sister company Stewart Milne Construction Ltd. Having weathered the local recession, the company is performing well and freely admits its success is due to its good local knowledge and a good working relationship with the district council as LPA.

Private house building in the Aberdeen area, as elsewhere in the UK, is predominantly undertaken by developers and building companies. Land is assembled by them, planning permission obtained and building completed on a speculative basis, though sales activity will commence shortly prior to work on site. House building by individuals is a small element of annual completions and is almost entirely confined to single units in rural locations or infill urban sites.

The second-hand housing market

In the Aberdeen area, over 90% of second-hand properties are sold through solicitors. This command of solicitors of the second-hand housing market, which extends into mortgage advice and mortgage broking, has some advantages, offering a one-stop service to users and lower sale costs, the estate agency commission fee being subsidized by the legal fees. Whereas in England the great majority of second-hand houses are invariably marketed through an estate agent, it has traditionally been the case in Scotland that this rôle is undertaken by solicitors, though numbers of estate agents are increasing.

To strengthen their position in the market, solicitors often form local associations operating a common property shop. In the case of Aberdeen, it is called the Aberdeen Solicitors' Property Centre and is conveniently located in a shop unit close to Union Street, the main shopping street. It issues a weekly bulletin of available properties, and details of all dwellings currently on the market in the area may be consulted there. This service supplements press advertisements by individual firms of solicitors and advertisement from their own often high street premises. Aggregated data from the centre also provide a very good guide to the market in the area.

Sales of second-hand residential property, as with new housing described above, are concluded with an exchange of missives, but this occurs at a much earlier stage in the transaction process than in England. Prospective purchasers submit bids "blind", i.e. without knowledge of any other bids being made, to the solicitor handling the sale. When a bid is accepted by the vendor, missives would normally be exchanged within 1–2 days. At this stage, the agreement to buy and sell is legally binding unless the vendor's title to the property is not proved or some other legal problem occurs. It is not possible for the purchaser to withdraw because of a change of mind or for a vendor to accept a later bid.

This contrasts with the English position, where exchange of contracts comes after title is proved and often some weeks after the offer to buy is made and accepted subject to contract. One consequence is that "For Sale" signs remain on display longer in England, as the sale may fail and/or better offers may be forthcoming. Also, whereas in England offers and actual selling prices are both often below advertised prices by as much as 10–20%, depending on market conditions, in the Aberdeen housing market successful bids leading to actual sales may be around 10% above the advertised price.

PART III
The urban property market

CHAPTER 7

The legal environment

7.1 Legislation

It is quite impossible to separate the operation of the Town and Country Planning Acts in respect of first urban use (i.e. the urban land market) from that in respect of subsequent uses (i.e. the urban property market), as the law is so written as to make no distinction. Change of use, as explained in Chapter 3.1, is subject to the same law, procedure and system of development control as first urban use or any subsequent development or redevelopment.

Historic buildings

Buildings and areas of special architectural and historic interest are protected under the Planning (Listed Buildings and Conservation Areas) Act 1990 (LBCA Act), which consolidates previous enactments and supplements the 1990 TCP Act. Ancient monuments (i.e. archaeological remains and ruins no longer capable of being occupied) are preserved under separate legislation, the Ancient Monuments and Archaeological Areas Act 1979.

Under the LBCA Act, the Secretary of State for the Environment is charged with the duty of maintaining a list of buildings of special architectural and historic interest under sections 1 and 2 of the Act. The criteria for listing a building are to be found in Circular 8/87 (Historic Buildings and Conservation Areas – Policy and Practice). On listing, buildings are classified in grades to show their relative importance. Surveys of areas are carried out to identify buildings justifying listing, but these are not necessarily comprehensive or exhaustive. Many buildings achieve listing only when threatened with redevelopment or demolition. The list includes buildings constructed as recently as the 1950s.

The demolition of a listed building or any alteration or the extension of the same in any manner likely to affect its character as a building of special architectural or historic interest requires the express consent of the local planning authority through an application for listed building consent. The local planning authority must advise the Secretary of State of the application, and he may call it in for his decision. The Historic Buildings and Monuments Commission (generally known as English Heritage), which assists the Secretary of State in determining listings, must also be notified.

These provisions extend to internal alterations, such as the removal of a fireplace, and to any object within the curtilage of the building such as entrance gates, and apply in addition to the development control provisions contained in the 1990 TCP Act.

If an unlisted building that might merit listing is in danger of demolition or alteration, a local planning authority can serve a Building Preservation Order under the LBCA Act. This stops any works until the question of listing can be considered. If a decision is subsequently made not to list then a claim for financial compensation can be made against the local planning authority under the LBCA Act.

The listing of a building does not in itself entitle an owner or occupier to compensation or to a right to be consulted on the matter. Notification of the listing must be given, however, and the listing registered in the Register of Local Land Charges. Failure to obtain listed building consent is a criminal offence and ignorance of the listing is not a defence. The only defence available is limited to the need to carry out urgent repairs and is narrowly defined.

Conservation Areas are defined by the LBCA Act as areas of special architectural or historic interest, the character and appearance of which it is desirable to preserve and enhance. The local planning authorities are charged with a duty to identify these, designate them as such and publish proposals for their preservation and enhancement. Most older town or village centres and many older residential neighbourhoods are designated as Conservation Areas.

Following designation, stricter planning controls operate: publicity by way of a site notice and advertisement in a local newspaper must be given to any application for planning permission that in the opinion of the local planning authority affects the character of the area; for other than minor developments the Historic Buildings and Monuments Commission must be notified; the permitted classes of development under the GDO are frequently restricted; all trees are protected; and demolition of any sizeable building is prohibited.

Condominiums/multi-occupancy buildings

English land law does not include any powers enabling the imposition of positive land obligations enforceable against successive proprietors, and provides only for negative covenants to restrict future use. Ownership through the freehold tenure of flatted properties or other multi-occupancy properties does exist but this is rare, and the legal arrangements to ensure, for example, that each proprietor is obliged to contribute to the maintenance of the common roof are complex and convoluted. Instead, long leases dominate, with common issues being controlled by the freeholder. This issue is now being addressed by new legislation (see Ch.1.1).

The position in Scotland is different. There the feudal system of land tenure and a tradition of building in "tenements" has produced a well developed common law encoded in judicial opinions.

Rent law

Only residential properties are subject to any statutory rent controls. In any rent review of non-residential property the manner and timing of the review will be dictated by the terms of the lease, as also will the manner of resolution of any dispute. It is normally the case that disputes are resolved by reference to the expert opinion of a chartered surveyor to be appointed by the parties, or, in the absence of agreement between them to agree such an appointment, by a surveyor appointed by the Royal Institute of Chartered Surveyors. There is no provision in law for statutory intervention.

Assured tenancies The rent for the original duration of the lease is a matter of agreement between the parties. On the expiry of the duration of the lease the tenant may nevertheless continue in occupation unless grounds for possession are established by the landlord (see Ch.1.1). The landlord may then serve notice under the Housing Act 1988 for a review of the rent. If the parties cannot agree then the rent is assessed by the Rent Assessment Committee, a government agency. The rent is assessed by reference to market rents prevailing in the area for similar properties. Reviews of the rent are possible at intervals of one year. If, however, a lease has written into it its own provisions for review of the rent then no resort to the Rent Assessment Committee can be made. The Rent Assessment Committee for each area must keep a register of determined rents for public inspection.

Assured shorthold tenancies At any time during the period of the lease a tenant under this form of lease may require a review of the rent by the Rent Assessment Committee by reference to prevailing open market rents. The Rent Assessment Committee can however intervene only if it considers there is a sufficient number of assured lets in the area to allow comparison, and the rent charged is significantly higher than the landlord could expect to get on the basis of these comparisons.

7.2 Possibilities of public intervention

The development control system discussed in detail in Chapter 3.1 applies equally to property as to land. In the context of the urban renewal of residential areas, certain legal instruments and policy designations apply specifically to the property market.

General Improvement Areas (GIAs)

Under the Housing Act 1969, local authorities were able to designate areas of older housing as GIAs, within which improvement grants were made available for individual dwellings, and environmental improvements (e.g. traffic calming, landscaping, provision of off-street play areas or car parking) were carried out by the local authority.

Areas designated typically included areas of better-quality housing where demolition was not justified, often built in the period 1875–1914 and originally in the private rental sector but often in recent years owner-occupied. GIA designation of older (1920s) council housing estates was also sometimes (particularly before the right to buy) adopted by local housing authorities as a measure for undertaking improvement of their own stock. GIA designation opened the way for building societies to offer mortgages on such older property, giving confidence that the area would not be subject to redevelopment within 30 years so allowing time for a normal repayment period of 25 years.

Housing Action Areas (HAAs)

This designation was established under the 1974 Housing Act to focus on areas where the dwellings were not necessarily conventional family houses, possibly in multi-occupancy, and where very low income households were to be found. Emphasis was more on social rather than property improvement, policy objectives in response to the process of gentrification that occurred in some GIAs. Again, confidence that the area would not be redeveloped facilitated availability of loans.

Housing Renewal Areas (HRAs)

Under the Local Government and Housing Act of 1989, GIAs and HAAs have been replaced by HRAs, although designations made previously remain and continue to be known by the original terminology.

Powers for intervention by local authorities to secure the improvement of a residential environment are contained in Parts VII and VIII of the Local Government and Housing Act 1989. Guidance on the terms of the Act is provided by the Department of the Environment in Circular 6/90. These powers are rarely used.

A local housing authority (i.e. a district, not county council) may designate an HRA if, following a detailed survey, it can identify a residential area encompassing a minimum of 300 dwelling-houses, or 500 dwelling-houses in an inner-city area, of which (a) 75% are unfit for human habitation as defined by statute or lack basic amenities; (b) 75% are privately owned; and (c) 30% of the households are dependent to a significant degree on state benefits.

Following designation, powers of compulsory purchase are conferred on the district council for the purpose of securing the repair and renewal of dwelling-houses either by the council or a third party nominated by them, such as a Housing Association. Compulsory purchase powers extend also to the improvement and provision of public amenities. Levels of grant assistance available under the Renovation Grant scheme also introduced by the 1989 Act and discussed below are raised to 75%.

Few designations of HRAs are expected given the stringent requirements for designation. The 1960s and 1970s saw mass clearances of private sector slums and the expense of a detailed survey discourages authorities from identifying potential HRAs areas in the knowledge that one or other of the criteria is unlikely to be met albeit that the area in question might reasonably be considered blighted.

CHAPTER 8

The demand for and supply of property

Information on trends in demand and supply has already been discussed in Chapters 1.4 and 1.5. Generally it is important to recognize that demand and supply in each property sector are tied to property cycles and that these have occurred at regular intervals during the post-1945 period. The continued existence of inflation, or the anticipation of inflation, will almost certainly ensure that the cycle will be repeated in due course. UK membership of the ERM was intended to mark a fundamental break with the past in this respect, but following a long period of economic depression the financial markets decided that UK interest rates had to be reduced and the markets forced the suspension of UK membership in September 1992. Interest rates were duly reduced and the economy is showing signs of improvement. In the commercial property market the reduction in interest rates has resulted in a substantial increase in investment demand.

8.1 Housing

Economic cycles and the persistence of inflation have encouraged the British to think of property as an investment. Even owner-occupied housing is commonly viewed in this way and owners regularly check the property press to see the asking price for property similar to their own. During the late 1980s the rise in house prices led to a general feeling of wellbeing in society and a willingness to borrow and spend despite the fact that very few owners attempted to realize their capital gains by moving to cheaper property.

Housing demand follows the economic cycle and plays a substantial rôle in reinforcing it. From the bottom of the cycle, housing demand increases as first-time buyers enter the market with the perception that house prices can only rise and that they should buy before that occurs. Existing owners also can choose to move and they commonly borrow more to finance a more expensive house. The market tends to accelerate over several years until a point is reached where house prices are rising very rapidly and people who choose to move have to be sure of finding another house very quickly before it becomes even more expensive. The cycle tends to peak as people recognize that prices cannot keep rising and the government increases interest rates.

After the peak of the cycle is reached, house prices stop rising and will fall, at least in real terms and recently in money terms, and it becomes difficult to move house as the level of transactions slows dramatically. The cycle is dominated by the demand for housing funded by mortgage finance. The cycle is not regular and therefore its length is not easily predictable. In addition, the heights to which the booms reach and the depths of the troughs are not regular.

Housing supply is also tied to the housing cycle because almost all new homes are purchased by mortgage finance. Housing supply is responsive to the cycle rather than a cause of it.

8.2 Commercial property

Demand

The demand by occupiers is clearly tied to the economic cycle. It has a spatial dimension, discussed elsewhere (Ch.1.5), and a structural dimension related to the de-industrialization of Britain and the move towards a service sector economy.

The demand by investors to purchase property is less tied to the economic cycle. Interest rates and the outlook for other investment mediums have an important effect on it. International investors are particularly important in some sectors (see Ch.9.1).

Supply

The supply of new properties for occupation is tied closely to the economic cycle because it is only in good trading conditions that firms are willing to occupy additional space. Land-owners additionally recognize the need to find an investor who is willing to purchase the property when it is complete, and therefore much development occurs when both the occupier demand and the investment demand are expected to be strong.

Finally it should be noted that commercial businesses can, by careful exploitation of their property assets, considerably increase their profitability. One result is that many attempts at financial take-overs via the Stock Exchange are based upon the intention to exploit property rather than on the obtaining of the business activity. One interesting example of this behaviour was the sale in 1990 of British Leyland by the government to British Aerospace. Much of the debate centred on the valuation of British Leyland's property assets, while the question of the appropriateness of British Aerospace as the owner of a car manufacturer received little comment.

CHAPTER 9

Case studies of the property market

9.1 London's office market

Background

London's office market has developed in the post-1945 era to become one of the world's largest concentrations of such property. Within the UK the absence of strong regional government and the growing dominance of national and international firms at the expense of "local" firms have facilitated the centralization of the "office function" within the nations capital city. Central government, itself a major occupier of office space, has attempted to relocate some of its activities, especially routine functions, into the regions but the predominant location of government offices remains central London.

Today there is a total stock of approximately 18M m², a considerable proportion of which has either been built or completely refurbished since 1985. The most prestigious locations are within an area known as "the City" and based around the capital's financial sector. A more diversified tenant base is found in the West End, and the remainder of the stock, occupied by a very wide selection of tenants, is found in an outer ring. Almost all of London's office market is, however, found within what would conventionally be called the "city centre" and the vast majority is north of the river.

In recent years the growth of the service sector in Britain, especially the banking and financial sectors, has reinforced the post-war trends, for both national and international financial activity is centred on London.

The geographical boundaries of the market are difficult to define. For this reason, this case study cannot be identified with any specific LPA. Many tenants who would prefer a very central location are "pushed" into less preferred locations by the very high city centre rent levels.

The geographical dominance of London can be illustrated by the fact that approximately 40% of Britain's office space is in Greater London and another 15% in the rest of South East England. These figures understate the true dominance of London because much of the office usage in other locations is either of a secondary nature (e.g. simple clerical tasks) or cannot be relocated away from the local market (e.g. accountants, solicitors, and other local professionals). In effect, the London office market houses the decision-making centres of both government and industry as well as being one of the global centres of the financial services industry.

The market is characterized by its division into a user market and an investment market by the widespread utilization of the leasehold tenure form (see Ch.1.1).

For the user, the market offers a variety of space. Units vary in rent, size, location and quality. For tenants requiring relatively small units, the choice is almost unlimited in normal circumstances. For occupiers of larger units the choice is more limited and, in conditions of strong demand, it may be very difficult to obtain large units quickly. Users have to negotiate with owners over rents and other elements of the lease and need professional advice from a firm of surveyors.

The investment market can function quite independently of the user market because the sale of an office building from one investor to another will not directly affect the tenant, though naturally the rents paid in the user market will affect the prices paid in the investment market. It is the dominance of the investment market that is unusual in European terms, and indeed the British perception of commercial property as an investment medium rather than as a capital item in the productive process should be emphasized.

During the post-1945 period, rents and capital values have normally been determined by the free play of market forces and taxes on capital gains and rental income have generally been set so as to avoid disadvantaging property as an investment medium. Government has influenced the market, however, via town planning policies that have influenced the property development process and, during the 1980s, by the creation of the London Docklands Development Corporation and an Enterprise Zone.

While commercial property is generally regarded as an illiquid asset by investors the London office market is far more liquid than that found in other European cities. There are many investors and a very large selection of properties. Buyers and sellers can always find a competitive market for property and a true market price.

For many years the market has offered good returns to investors, and this and the lack of government intervention, coupled with capital mobility, have made the London office market attractive to international investors.

In trying to understand London's office market it is first necessary to understand the elements of the market: tenure, investment, actors, information and property development.

Tenure While owner-occupation exists, the London office market is dominated in terms of rents and capital values by the leasehold tenure. This has been explained in Chapter 1.1 but a few additional points need emphasizing.

The typical lease, dominant in the market until very recently, is one of 25 years where the tenant is responsible for all insurance and repairs to the property and the rents are reviewed (upwards only) to current market levels at regular intervals (5 years is common). The rent review is open to negotiation, but in the absence of agreement the case must go to arbitration, the results of which are binding on both parties. The tenant cannot evade the obligations of the lease except by either bankruptcy or the assignment of the lease to another occupier who is willing to accept the terms of the lease. Landlords may be willing to "buy" the lease from the tenant though usually at a negative price because the landlord is forfeiting the obligations of the tenant to pay rent for the remainder of the lease.

This leasing structure, which is highly advantageous to investors, has been maintained in the London office market because the generally high levels of demand have ensured that property owners have been in the strongest bargaining position. International investors have been particularly attracted to London because of the reduced risk that long leases offer.

In addition to the rent, tenants are responsible for paying a service charge for any services provided by the owner and the Uniform Business Rate.

Investment Property investment has in the past been undertaken by a variety of organizations: British pension funds, insurance companies and property companies, as well as financial institutions from overseas. The form of British investment in property has been strongly influenced by the British personal tax structure, which has encouraged individuals to invest via insurance policies and pension funds rather than via ownership of equities. While this changed somewhat in the 1980s with the government encouraging direct and indirect equity investments by individuals, the importance of the financial institutions has remained. The result has been that the insurance companies and pension funds have become the dominant owners in the London office market.

Investment in office property in London is commonly seen as comparable to investing in company shares or government securities. The investment is perceived as long term because property sales take time to complete and there are additional market problems of non-homogeneity and bulkiness.

Investors look to long-term performance and are prepared to accept low returns in the short run in exchange for the higher returns that rental growth will bring in the future. Valuation methods are complex, though a common method is via the yield (see Ch.4.1). Knowledge of obtainable rents and yields from the sale of similar properties enables valuers to estimate the value of each property. Over time as exchanges take place the yield will change with market conditions, and revaluation of other properties will be possible without their sale.

It is important when considering the functioning of the London office market to examine why it is dominated by the investment market rather than the tenure of owner-occupation. There is no single explanation of this practice; rather a number of financial pressures, laws and cultural practices have come together to promote the investment market.

The first issue is that ownership of property ensures that the returns from the eventual redevelopment of the site will be obtained. As office development in London has been financially very successful, investors have been willing to pay more now for this eventual return than owner-occupiers, who are generally less concerned with redevelopment opportunities.

Secondly, in general planning authorities in London have been restrictive, not permitting the creation of large quantities of new office space. During the 1980s this policy was changed, but for much of the post-war period there was a shortage of space in the city. Rents have been driven up, generating good returns for investors.

Thirdly, inflation in Britain has been an important issue since 1945, and the real asset nature of property has ensured that it has some degree of inflation resistance. This is another reason for investment by financial institutions.

Fourthly, the rôle of London as a major financial centre has ensured that many buildings are occupied by non-British companies. These firms are more likely to rent property than own it because they do not wish to invest the considerable sums that ownership would entail. Any loss of London's status as one of the world's top three major financial centres (alongside New York and Tokyo) to another European city could significantly affect this.

Fifthly, the landlord and tenant legislation clearly sets out the form of relationship between landlord and tenant. The leases created under this legislation have been tested in the courts and today clear legal advice can be obtained on the meaning of each clause. Investors should be fully aware of their rights and obligations.

Sixthly, the commercial companies whose shares are quoted on the stock exchange would commonly expect to be able to use their capital to better effect within their own business rather than via ownership of property. Interest rates have often been high in Britain and this has put additional pressure on companies to obtain the maximum return from their own financial resources.

Seventhly, property ownership for a company is not a necessary requirement before the banking system will lend money for business development. Banks can and do lend to companies on the basis of their potential rather than on the basis of their assets. Interestingly, property developers are one of the beneficiaries of this situation because there are numerous examples of the banking system lending to finance what the bank sees as a viable development opportunity despite the company having few assets.

Eighthly, it would be wrong to ignore the effect of culture. Britain's office market has been dominated by the investment regime for many years. Both owners and occupiers expect this tenure, and professional services (most notably surveyors) develop to facilitate the functioning of the market. It will take the arrival of some outside pressure such as new occupiers with different traditions to change the status quo.

The state of the investment market is heavily dependent on the general outlook for the economy as a whole and for future property demand in particular. The lack of stability in the UK economy has been reflected in the property market, and the concept of a property cycle is well understood if not well predicted.

Actors and information flows Given that each property is unique and that there are a large number of properties within the London market, users and investors would find the market highly inefficient if it were not for the existence of an effective information system.

Information is not centrally collected by either a government department or private agency but rather groups of property professionals, particularly surveyors, compete to offer their services to clients. Their most valuable resource is their knowledge of local market conditions and consequently they undertake much research on the local market.

The following list of information gives some idea of the data collection that is undertaken:

(a) Market information available on a spatial basis: rents, capital values, yields, rental growth, capital growth, vacant space, new lets, inward and outward mobility, property sizes.

(b) Occupier information: types of tenants, space requirements (air conditioning, etc.), locational needs, employee characteristics, users of information technology.

(c) Investor information: investment by type of investor, international investors by nationality and type of purchase, investor requirements (size, location, etc.), investment portfolio balances.

Although contractual relationships between landlord and tenant are technically private, there is little doubt that the information on key issues such as rents, capital values, etc. is widely available within the surveying profession. Information flows are enhanced by the fact that the property professionals are all members of the same professional body and that young members of the profession commonly move from firm to firm.

Overall, the market professionals act so as to ensure that information is readily available and that the market can efficiently respond to changes in occupiers' and investors' demands. Additional research has been undertaken within the academic community and within the planning profession, who are commonly concerned about the planning consequences of property development activity (Marriot 1989, Thornley 1990, Healey et al. 1992).

Property development The functioning of the London office market cannot be fully understood without an appreciation of the property development process. Developers build properties speculatively when they perceive a profitable opportunity. Offices are normally built by development companies and either held by the developer as an investment or sold on to an investing organization. Either way the investment value is a crucial issue, for it and the costs of construction determine the profitability of development. When yields are high because of a lack of investment demand, the value of properties is low and development is less likely to be profitable. The explanation of high yields is that low investment demand reflects a lack of occupier demand and consequently a low expectation of rental growth.

The rate of property development is controlled by the planning system. Within London in the post-1945 era there has been a tremendous growth in the level of office building but, as was noted earlier, the process has been cyclical. Equally important is the fact that up to at least the beginning of the 1980s the planning system had generally been somewhat restrictive towards office development in London. Considerable effort was made to stop the demolition of old buildings and there was a view that developers should be encouraged to build outside the city centre and if possible outside South East England. From the developers' point of view, the high rents in the city centre encouraged central development, but high land prices and the difficulty of obtaining large sites from numerous existing owners have always made the development process difficult.

In the 1980s the government's pro-market and pro-property development policies in effect ensured that property development activity was not unduly restricted by the town planning system. Indeed, some London boroughs tried to exploit the high levels of developer demand to obtain planning gain. With very high levels of demand for property from both users and investors, the 1980s saw rapid development activity, which reached a peak in new starts in 1989. This has been delivered as new office space during the current slump, as the figures in Table 9.1 indicate.

The government intervened in the 1980s with the creation of the London Docklands Development Corporation which has attempted to redevelop previously derelict land via compulsory purchase, land improvement, the provision of infrastructure and direct subsidies. While the buildings it has created form an important element in the total level of property development (435,000 m² of office space) they have had only a modest effect on the investment market, which covers a much larger quantity of space.

Property development in the future will be possible only if there continues to be a growth in demand to be located in the city. Some decentralization has occurred as businesses have attempted to utilize the lower rents and labour costs in other locations, but this pressure has been somewhat defused by the current surplus of space and correspondingly lower rents. Current evidence (see *Estates Times* 18/9/199) suggests that the growth in employment in the city will be slow during the rest of the 1990s.

A more serious long-term problem may be the increasing level of congestion within the city, which has not been addressed by investment in transport infrastructure.

Current market conditions

Today the market is so extensive that owners know that they can always find actors willing to buy their premises at the going market rate and that even in a depressed market there will be new occupiers intending to locate in the city. Additionally, some existing occupiers will be intending to relocate as their property requirements alter or their lease comes to an end.

Since 1989 the market has moved into a severe depression, but the stock has continued to grow as developments have been completed (see Table 9.1). Currently development has almost ceased and this has led to worries that the stock of new property may be inadequate at some future date.

Table 9.1 Office development completions in London, 1992–5, ('000 m²).

Year	West End	City	Docklands	total
1992	164	242	96	502
1993	145	84	0	229
1994	44	13	0	57
1995	53	15	0	68

Source: Savills.

Rental growth was very rapid at the end of the 1980s, as Table 9.2 indicates, but has suffered with the recession. Office rents have fallen from their peak by over 50% in prime locations. The prime rents at the peak stood at over £700/m² and were the highest in the world outside Japan. The yields on these prime buildings fell as low as 5.25%, resulting in £100 rental income costing a capital value of £1905. By mid-1992 London was no longer such an expensive city in which to locate (see Table 9.3).

165

Table 9.2 Rental growth in London at the end of the 1980s.

Year	%
1986	14.2
1987	32.2
1988	29.4
1989	16.8

Source: Healey and Baker.

Table 9.3 Comparative office rent levels in mid-1992, ($£ m^2$).

City	Rent level
Paris	409
Frankfurt	377
Berlin	377
Madrid	334
Munich	312
Milan	312
London	269
Brussels	172
Amsterdam	140

Source: Jones Lang Wootton.
Note: All the figures include appropriate allowances for rent-free periods and other "discounts".

The property slump has had a number of unexpected consequences. First, the rent levels in areas that are regarded as not prime (central) have fallen to very low levels and properties are almost impossible to let, with a vacancy rate of over 30%. Whether these properties will ever be let is debatable and certainly older and less well equipped premises will need extensive modification before occupation. Some redevelopment for housing may occur. Valuation of these properties is difficult.

Secondly, many occupiers were forced to pay high rents under the terms of their lease (upward-only rent reviews) and now cannot sublet their property if they wish to move except at much lower rents.

Thirdly, property owners are increasingly offering inducements to occupiers, such as rent-free periods and shorter leases with break clauses.

Lastly, the Docklands development, once seen as the most prestigious development in Europe, has a vacancy rate of at least 50%. The main developer, Olympia and York, has collapsed and rents are as low as $£0.92/m^2$. The development is seen as inaccessible because of the failure to address its transport requirements.

After the exit from the ERM in September 1992, and the corresponding fall in interest rates, there has been a renewed interest in investment in the London's office market. Much of this demand in the first part of 1993 came from overseas, where property yields are quite low – Germany 5%, France 6.5%, Spain 7%, compared with 8–9% in Britain. As the economic recession appears to be ending, these investors take the view that the current property values will not fall further and are now "cheap". The next step in the cycle appears to have been reached!

166

9.2 Radlett and Shenley, Hertfordshire, housing market

This case study is of the residential property market in and around the small town of Radlett, in Hertsmere district in the county of Hertfordshire, north of London.

It was selected as a typical example of an area with high property prices in the Outer South East. It specifically illustrates the property market because very limited amounts of land are available for new housing and there is virtually none on greenfield sites owing to strict green belt and other planning policies.

Geographical and planning context

Hertfordshire contains a large number of distinct urban communities, including the substantial commercial centre of Watford, the historic cathedral town of St Albans, a number of the early new towns, including Letchworth (1903), Hemel Hempstead and Stevenage (both part of the post-war ring of new towns around London), and settlements such as Borehamwood, which originated as an overspill estate developed by the former London County Council in the 1950s and 1960s to rehouse people displaced by slum clearance in inner London.

The county council has for many years maintained a strict green belt policy in order to ensure that these urban settlements do not coalesce and to retain some open countryside. It also has the reputation for formulating coherent planning policies and for following them. It is the only county in England within which all district councils prepared district plans covering the whole of their territory during the 1980s, before this became a government requirement.

Hertsmere district was a new creation following local government reorganization in 1974. It includes the towns of Borehamwood (where its administrative offices are located), Elstree (once home of the British film industry), Potters Bar and Bushey, and is crossed by the M25 London orbital motorway. Radlett is not regarded in the structure plan as a town, it is defined as a "specified settlement" (see below). Much of the non-urbanized part of its area is within the metropolitan green belt, apart from the site of Shenley Hospital.

Hertsmere's population is estimated at 88,600, of whom 43,300 are economically active. Of these, 70% commute to employment outside the borough. Population growth is concentrated in the under 15 and elderly groups, with decline especially in the 15–24 group. Of households, 33% own two or more cars and 68% are owner-occupiers (Hertsmere 1990). The average house price in 1989 was £93,500, or 44% above the national average. This disparity is now substantially less. The number of planning applications determined by Hertsmere in 1989 fell by 2% to 1450 after several years when the numbers increased, reflecting the beginning of the downturn in the property market. Politically the Conservatives control Hertsmere and have traditionally predominated throughout the county, although they lost overall control in the 1993 county council elections. The previous Conservative hegemony did not, however, imply rejection of planning controls over development in pursuit of a free market ideology.

Radlett has a population of about 8000, and is located 5 km north of Borehamwood, 7 km south of St Albans (the M25 running between the two) and about 26 km from central London. It is served by an electrified commuter rail link to London Kings Cross (journey time 20–25 minutes) and also by the cross-London Thameslink line to the city and the rail network south of London. It is basically a residential community with its own identity, physically separate from other settlements and offering very convenient access to London's employment market.

Housing

Housing mostly consists of detached family houses of three to six bedrooms built at different periods from the beginning of the 20th century until the 1960s. Overall, Radlett has a high proportion of large houses: 36% have seven or more habitable rooms, compared with 15% in Hertsmere as a whole; the proportion with six rooms corresponds to the district average and the proportion with five or fewer remains below average.

A total of 195 dwellings were completed in Radlett during the period 1981–91, in about 70 separate developments. Several of these were houses of four or five bedrooms on single unit plots; many of the remainder were of one or two bedrooms to balance local provision and meet a need for retirement homes. Traditional three-bedroom, five habitable room housing, common in new developments elsewhere, were rare in Radlett. There are very few sites on which proposals for new housing developments are likely to receive planning permission.

Planning policy

The county structure plan current at the time of the study is the 1986 review, which was formally approved by the SOS and came into effect in May 1988. In January 1991, revisions were put forward that were subject to an EIP later in 1991. This does not propose any changes of significance in relation to this case study.

The 1986 review includes Radlett in the list of "specified settlements" indicated on the key diagram, for which surrounding green belt boundaries are defined in local plans, and the policy is that "development will be limited to that which is compatible with the maintenance and enhancement of their character and the maintenance of green belt boundaries" (Hertfordshire Structure Plan 1986, Policy 50).

The Hertfordshire green belt is widely respected by developers, and it is generally accepted that any proposal to breach the green belt would be rejected by the local planning authority. Hertsmere has recently produced a new local plan, the Hertsmere District Plan First Review, which was published in July 1989, had its PLI in June 1990 and was formally adopted with effect from 1 May 1991.

Within the urban limits of Radlett, the only new housing possibilities are on small infill sites, or within the curtilage of existing dwellings. Such sites are often difficult to develop. The local planning authority will not normally grant permission unless the site proposed offers separate road access for each dwelling, and a satisfactory minimum distance is maintained between dwellings and from the edge of the site. A maximum density of 12 dwellings per hectare is also required under the district plan, along with an environmental appraisal of the site to ensure that the existing character of the area is maintained. Because development sites are so small, Radlett is not of interest to the bigger-volume house-builders, and the developers are entirely small local firms.

New planning guidelines were prepared in 1991 by the LPA for proposals of this sort because high property prices and shortages of sites for new building had led to the phenomenon of "town cramming" in several locations in the Outer South East, and a number of single dwellings had been granted permission in Radlett on appeal. The District Plan and guidelines were intended to avoid this problem by establishing firm policies and indicating what would be acceptable. Where an existing building is demolished, any new building on the site is expected to be within the same building print, or have no greater plot coverage, than the original building. Even with this restriction, market conditions in the late 1980s were such that it could be profitable to demolish an existing dwelling and build a new one on the same site. However, since 1991, pressure for development has eased with the fall in property prices and the incidence of cases of owner-occupiers with negative equity or having difficulty selling, so the new guidelines have not been tested as much as anticipated. The LPA has also undertaken a survey of open land in urban areas, as a basis for policies of development control designed to ensure that adequate open land is retained for schools, playing fields, recreation, etc.

Shenley

The site of Shenley Hospital is the only substantial site for new housing in the area. This is a former long-stay psychiatric hospital being offered for sale by the Regional Health Authority (RHA) in order to help finance its community care programme. The sale is being held up because of the fall in land and property prices, as the RHA must obtain the best possible price. The site is identified in the structure plan as a major housing site, with outline planning permission for a development of around 900 dwellings, phased over 15 years, on a site of 30.6 ha. A total of 18% of the housing land will be allocated for social (i.e. subsidized) dwellings for rent or purchase, on which 25–30% of the total number of units will be built at higher densities than the remaining housing. A planning brief for the development of Shenley commissioned by the borough council was prepared by planning consultants and published in November 1986. It makes detailed proposals for the development, including housing, shops, community facilities, school, small business accommodation, and parkland and recreation areas. The whole development would be the subject of section 106 planning agreements with the RHA concerning provision and funding of these components (i.e. planning gain).

The Shenley development is one of the biggest housing land allocations likely to come available in this part of Hertfordshire, and therefore will attract the interest of the big-volume house-builders who are not interested in the small schemes otherwise possible in places such as Radlett. The scale of the Shenley proposal is such that it is in fact regarded as a planned site for a private new town. The proposal is seen as strengthening the green belt by making a realistic and defensible allocation of land for new housing in a suitable location, thereby enabling the green belt elsewhere in the authority to be sustained. It therefore fits in the South East regional strategy concerning the planned release of housing land to meet demand where this is possible consistent with green belt and other planning policies.

Housing market

The remarks in this section are based on interviews with local estate agents. There was general agreement that the traditional demand for large family houses still prevailed, but that prices had to be set "realistically". Demand is coming from people wanting to move in to the area, especially aspiring Radlett residents seeking to move up market from Borehamwood, where values are substantially lower. There is also demand from new households formed by local residents and from older retired local people who no longer want a large house. The latter demand is largely frustrated; hence the limited emphasis on smaller dwellings among recent new construction. One national builder felt there is a huge potential for retirement homes and is undertaking a scheme of one- and two-bedroom flats, with a warden, near the shops. Nevertheless, in circumstances where only small developments are possible, profit margins on the development of smaller dwellings is less than on large family houses.

There is general acceptance among the estate agents that the LPA will not allow higher-density development on available sites, and that the green belt and other planning restrictions will not and should not be breached. A further consideration is the view that the prestige and reputation of the area, and therefore the sustainability of price levels, depend on its large houses.

Inward migration to the UK in the pre-war period is reflected in the local housing market: property in Radlett is advertised in the local weekly newspapers and also in the national *Jewish Chronicle*. The adjacent part of north London has historically been an area in which many Jews live, including many Jewish immigrants from central Europe, and this community remains strongly represented in this area as it spreads out from London into the commuting suburbs in the same geographical sector.

Typical prices

Prices at which residential property is currently advertised range from, at the lower end: £210,000 for a three-bedroom bungalow, £189,000 for a three-bedroom semi-detached house built in the 1950s, £250,000 for a four-bedroom 1930s house, up to £685,000 for a six-bedroom house in extensive gardens. Realistic pricing policies mean recognition that prices have fallen since the 1988 peak by about 20–30%, and during 1990 by over 10%. A slight decline in prices is continuing. Houses can still be sold readily; there is still a market, but recession and reaction to overheating in the market in the Outer South East in the late 1980s are causing problems.

Many purchasers of houses in Radlett at times of higher prices are finding that if they want to sell they cannot do so at the price they paid, or at a price that will cover the outstanding debt on the mortgage. Some are hanging on in the hope of achieving such a price (an example was quoted by a local agent of a house valued by them in mid-1990 at a "realistic" price of £350,000, advertised by another agent at the price the vendors were hoping for of £450,000, and sold after one year at £340,000). Others are forced to sell to cut their losses, as they can no longer serve their mortgages. An additional complication is that some people took on non-sterling mortgages in order to avoid high UK interest rates.

Dwellings built in the 1950s, 1930s or earlier hold their value well if in good condition, as do the few 1980s dwellings available. Newly constructed dwellings also sell well, but come on to the market at a slow rate. The Nightingale Close development by Bishopwood Estates, for example, was described as selling well. A total of 15 detached houses are being built and were originally expected to sell at around £550,000 each, but became available in the £300,000–400,000 range. Six were already sold, and all were expected to be sold by the end of 1992.

Other factors

Recession is hitting the area and forcing a number of house-owners with very large mortgages to sell. Unlike the recession of the early 1980s, the current recession is affecting many employed in professional services and commerce in the South East. Radlett is especially affected by redundancies among accountants and other employees in the financial services industry in central London.

Realism is being forced on the market not only by actual sale prices achieved but also by mortgage valuations being offered by chartered surveyors on behalf of building societies.

The market in Radlett is largely in existing property in an area tightly constrained by planning policies and therefore not subject to large numbers of new dwellings coming on to the market, at least until Shenley is developed. Price fluctuations therefore reflect market conditions as they affect the section of the population economically able to consider living there.

9.3 Falstone Walk, Fawdon, Newcastle upon Tyne: renewal of social housing

This case study examines a partnership between the public and private sectors, actively encouraged by central government through grant assistance, for the provision of low-cost owner-occupied housing and the regeneration of a residential environment.

The study site lies approximately 5 km northwest of the city centre and is within the City of Newcastle upon Tyne District. It is a local authority housing estate built in the mid-1950s comprising at the outset of the project 11 five-storey blocks ("the blocks"), each containing 20 one-bedroom units, and 105 two-storey, semi-detached and terraced two- or three-bedroom houses ("the houses").

The blocks had a history of design failure. They lay in an unbroken line to the west of Dorrington Road with drab exteriors, flat roofs and an absence of any landscaping. This conveyed the impression of a bleak imposing and unbroken wall. They were of solid concrete construction and in consequence had poor insulating qualities. Originally, they were internally designed with one-bedroom accommodation at ground and first-floor level for the elderly or disabled, and three-bedroom maisonette accommodation for families above, accessed at second- and fourth-floor levels by internal staircases. By the early 1970s the maisonettes, inherently unsuitable as family accommodation, were proving hard to let and vacancies were high, bringing associated problems of vandalism, lack of security and social stigma to remaining residents.

In 1975 the city council sought to address this problem by the internal conversion of each block into 20 one-bedroom flats. This was promoted on the basis of an acute shortage in the public sector at that time of single-person accommodation. The density of 20 units per block (220 units in all) and the high concentration of young people in relative isolation to the city centre and services proved to be a disaster. By the early 1980s the problems of vacancies, vandalism, etc. had returned. In August 1988 only 86 of the 220 units were occupied.

The houses lying to the east of Dorrington Road in an irregular grid pattern, by contrast, proved reasonably popular with tenants, albeit that they shared the same insulation problems, being of solid concrete construction. They enjoyed a low density and sizeable garden grounds. That popularity was, however, waning as the problems of the blocks worsened and a general decline in the estate was becoming apparent to all.

Central government constraints on local government expenditure on housing made it impossible for the city council to address the problems of the estate from its own resources. Its capital housing programme in 1979–80 had a budget of £22M; in 1990–1 it was £23.28M, a reduction in real terms of 50%. A survey in the late 1980s of its housing stock identified that expenditure of £170M was required to bring it up to acceptable standards. This reduction in local government resources stems from central government's gradual restriction through the 1980s of the local authority's ability to borrow to fund housing investment. Further borrowings are possible only with the approval of the DoE under its Estate Action scheme. That approval must be applied for for specific housing estates in need of regeneration, and priority is given to LA schemes that attract private investment and produce a diversification of tenure.

In late 1987 the city council discussed the possibility of Estate Action for Dorrington Road with the DoE, proposing the sale of five of the blocks with the remainder being improved by the council under Estate Action for rent. This was rejected by the DoE, but from the discussions the city council was given to understand that the sale of all the blocks to the private sector could attract Estate Action for the 105 houses. On this basis the council turned to the private sector, identifying Wimpey Homes Holdings plc as a possible partner to achieve this end.

Wimpey Homes Holdings plc had from the early 1980s been diversifying its residential construction activities away from reliance on greenfield locations to inner-city brownfield sites and conversion and refurbishment of commercial and other properties for residential use. By the late 1980s, encouraged by central government subsidy through City Grant and its predecessors, it had become involved in numerous partnerships with local authorities, acquiring "problem housing" from them and undertaking its

173

refurbishment for sale. This complied with government policy to extend owner-occupation to lower wage-earners. In the summer of 1988 Wimpey completed a successful partnership with the city council at Tyne Dale Rise, Throckley, to the east of Newcastle under which 20 houses, 20 flats and 4 new build houses were provided from a similarly dilapidated council housing estate for sale to lower-income groups. On the strength of that successful partnership Wimpey was asked to put forward proposals for Dorrington Road.

The initial approach to Wimpey by the city council was informal. No other house-building or construction company was approached. The broad understanding was that the blocks would be conveyed to Wimpey for extensive refurbishment and sale by it under the City Grant scheme and that the council would undertake the external refurbishment of the 105 houses under the Estate Action scheme. Wimpey agreed to prepare a scheme for the blocks. The city council appraised the DoE of this and received agreement in principle for Estate Action.

The location of the blocks dictated their redevelopment to provide low-cost housing for sale. The new Newcastle western by-pass was to run immediately to the west of the blocks, but access to the blocks would remain through a shabby mixture of private and public sector housing, despite the council's improvements. There was therefore no question of gentrification into an upmarket residential use.

To prepare a scheme, Wimpey employed the services of Jane Derbyshire Associates, a Newcastle firm of architects. Wimpey has emphasized their professional design skills as being a key element in the success of the project. There are two elements to the design: the visual breakdown of the "wall" through the reduction of five blocks to three storeys and the addition of external detail to all blocks, and the diversification of available units, including terraced dwelling-houses, through the demolition of one block at the entrance to Dorrington Road. In all, 110 housing units were provided, comprising a mixture of two-bedroom terraced dwelling-houses, three-bedroom maisonettes, three-bedroom flats, two-bedroom flats, and one-bedroom flats.

Following the presentation to the housing committee of the city council, the parties become involved in a complicated series of interlinked discussions and activities.

Planning permission Wimpey, through its architects, submitted an application for full planning permission to the planning department of the city council. No difficulties were anticipated or encountered in that regard.

City Grant From its previous activities, Wimpey had a good working relationship with the regional office of the DoE in Newcastle. On agreeing to prepare the scheme, the possibility of City Grant had been raised and positive signals given by the DOE. A formal application was submitted and the process of assessing construction costs and the value of the project was undertaken in order to assess the level of City Grant assistance that would allow Wimpey "reasonable" profit on its activities. City Grant would, however, be formally approved only on the grant of planning permission. The viability of the project in Wimpey's eyes was also dependent on the city council receiving Estate Action funding for the refurbishment of the 105 houses.

City Grant of £1.1M was eventually approved, a subsidy of £10,000 for each housing unit. A formal agreement was entered into that mirrored that agreed with the council (see below), with additional provision to allow the DOE to claw back 50% of profits in excess of those agreed with Wimpey.

Estate Action funding The city council resumed its prior negotiations with the DOE. The proposed works on the 105 houses were the addition of a second skin to improve insulation, new doors and windows, overhaul of roofs and chimneys, external canopy over entrance doors and new bin stores to the rear of the houses. Total costs of £1.2M were agreed with the DOE, 50% of which would be funded from approved borrowings under the Estate Action approval, the remainder from the city council's annual budget. The Estate Action element was, however, dependent on the capital investment of Wimpey in the blocks aided by the City Grant.

Partnership agreement The city council and Wimpey required to formalize their arrangement and the following agreement was reached:
(a) the city council must secure Estate Action grant from the DOE for the external refurbishment of the 105 houses;
(b) the blocks would be refurbished in accordance with the approved scheme;
(c) the blocks would be transferred to Wimpey at their market value; this is a statutory requirement, and the value was subsequently assessed at and approved by the DOE at zero;
(d) Wimpey would occupy the site under a licence and never receive title itself.

On completion of the works, the city council would grant a lease of 99 years and a day to Cheviot Housing Association, which would immediately grant a lease of 99 years to purchasers. The occupation under licence was agreed owing to its tax efficiency. Wimpey would be carrying out works of "repair" and these would be chargeable to VAT. As the council's "contractor" under the licence, VAT would not be chargeable, the council being exempt by its status from that tax. For the council's part, the retention of the freehold until completion also gave it an insurance against unforeseen disasters, such as Wimpey going into liquidation or receivership. The lease arrangements are standard practice given the inability of a freehold tenure to impose positive obligations such as contributions towards maintenance of common parts. The housing association would act as management agents, ensuring good future management of the blocks.

On release of the units, Wimpey would give priority to existing tenants of the council and those on its waiting lists or those of housing associations. This benefited both parties. Prospective purchasers would be from low-income groups in any event and the council would be assisting in the marketing process by identifying purchasers.

Matters began to fall into place in autumn 1989, starting with the grant of planning permission. There followed the award of City Grant and the approval of the funding under Estate Action.

Wimpey proceeded with obtaining a building warrant from the council and a road closure order from the Department of Transport. The scheme for the blocks required the removal of the spur roads that intersected them. These proved to be public highways and therefore formal road closure orders were required. There were considerable delays in obtaining these owing to inaction on the part of the Department. It appears that the requirement not to start work before receipt of the formal road closure orders was ignored, agreement being there but administrative delays preventing action.

The refurbishment of the blocks was undertaken in stages, starting with the demolition of the block at the entrance and the construction of the dwelling-houses, and progressing down the line of blocks. By August 1990 the first units were available for release. The units are spacious in comparison with other flatted developments and completed to a relatively high standard internally.

The prices were as follows: one-bedroom flat £22,750; two-bedroom flat £24,250 (later increased to £25,500); three-bedroom flats and maisonettes £29,500 (later increased to £30,950); and two-bedroom terraced dwellings £31,750 to £34,750, depending on position in the terrace and garden ground. The flatted accommodation was sold with vinyl floor coverings to wet areas and carpets to the lower floors. Washer dryers were also provided for upper-floor units. Ground-floor units were assigned garden ground and an adjacent parking space. Parking for other flatted units was unassigned but there was ample provision.

The council publicized the units through its local housing offices and mailshots to those on its waiting list. Wimpey undertook its usual marketing through the press and its showroom. Those entitled to priority under the parties' agreement were given three weeks' initial priority in making reservations. Demand has been healthy and purchasers have been forthcoming as units are completed, with some purchasers committed well in excess of anticipated completion dates.

On completion of the sales Wimpey proposed to do some research into the profile of the purchasers. Its view was that the purchasers would be predominately young childless couples and single persons from the lower-income groups who have some existing ties with the locality.

Table 9.4 Falstone Walk: timetable of events.

Late 1987	City council approach DOE for Estate Action
Aug 1988	First approach by city council to Wimpey, inviting preparation of scheme for blocks
Jan 1988	Agreement in principle by DOE for Estate Action
1st Apr 1989	Presentation by Wimpey to City Council of prepared scheme
24th Apr 1989	Planning application submitted in full
May 1989	City Grant application submitted after preliminary discussions with DOE
24th Sept 1989	Planning permission given by City Council.
24th Nov 1989	City Grant awarded by DOE
Nov 1989	Road closure order sought from Department of Transport
Dec 1989	Estate Action approved Work on site commenced by Wimpey
May 1990	Road closure order confirmed by Department of Transport Estate action works commenced
Aug 1990	Official opening for sale
June 1991	Estate Action works completed
1992	Sales of blocks completed

Source: Local authority records.

Summary

This project has been extremely successful in regenerating a residential environment. The 105 houses have been improved and the blocks successfully redeveloped to produce attractive housing. The wisdom of encouraging owner-occupation in lower-income groups is debatable, but central government policy implemented through the grant regime requires this. It is clear from the study that co-operation and mutual trust between the council and Wimpey were important, as good communication and liaison were critical in the negotiations. The DOE gears itself to be "user friendly" but there was criticism from both Wimpey and the city council of the amount of administration it generated.

An unanticipated bonus of the scheme became apparent with the opening of the new Newcastle western by-pass, which forms part of the London–Edinburgh trunk road (A1), in December 1990. The site had previously not been easily seen from anywhere other than the immediate neighbourhood, but from the new road it occupies a prominent and highly visible location. Without refurbishment the dwellings would have confirmed the worst stereotyped image of northeast England, but with the improvement works they create a much more positive impression.

PART IV
Conclusions

CHAPTER 10
Evaluation

10.1 Evaluation of the functioning of the urban land market

The main points to emphasize in the British urban land market are the discretionary planning system and private land assembly. These tend to give rise to the following features or characteristics: land banking; flexibility; development outside the plan; speculation; and negotiation.

Generally the British system exhibits a discretionary and policy-driven planning process that offers flexibility and negotiation to developers. The pressures faced by the planning system in the last 15 years, during which time different forms of development plan have been progressively introduced, along with considerable variations in government policy towards freedom of development and the standing of LPA plans and policies, have demonstrated a certain robustness in the system. The basic structure of the planning system has shown itself capable of accommodating and surviving very different political approaches to planning.

The British system does, however, have certain clear drawbacks. First, there is a lack of certainty. A considerable amount of land is developed that has not been identified for development in the planning process. Although such developments are profitable for the land-owners and may be socially desirable, there are clear questions of democratic accountability and externality effects on adjacent users. The process also makes land valuation very uncertain, and can undermine development appraisals because a developer may not have any certainty that the local authority will not permit another competing development in the locality.

Secondly, the negotiation of development permission may involve the local authority agreeing to development that it would not normally permit on planning grounds because it is offered benefits by the developer. This planning gain may be beneficial to the community but it is clear that the process is open to abuse.

Thirdly, the limited rôle of the state in the process of land assembly means

180

that the private sector faces the burden of this task without compulsory purchase powers. Developers often take many years to acquire all of the legal interests in a site, during which time the area may fall into decay. In addition, developers have an incentive to engage in land banking, seeing this both as a source of profit whenever planning permission is granted and as the method by which they can assure themselves of a supply of suitable land upon which to develop in the future.

Fourthly, it is very difficult to co-ordinate provision of infrastructure by the public sector with the development process. Accordingly, there is an additional lack of certainty regarding the provision, or timing of provision, of any new infrastructure that may be required. The result is that developments tend to be built near existing infrastructure, causing congestion and "over-development". For internationally mobile development, this factor, more than any feature of the planning system and land and property markets as such, is likely to influence decisions on whether or not development occurs in the UK or elsewhere in the EC.

Housing

Concern about the provision of housing land was the motivation for the first UK planning legislation (the Housing, Town Planning, etc. Act 1909) and has been one of the long-running issues in British planning. Land banking on the part of the housing developers, especially the bigger operators, is seen as crucial to their commercial success in a fluctuating property market.

Land is allocated in all local plans and there is a strong public policy commitment to ensuring that sufficient land is available for the private housing market. Nevertheless, house-builders still argue that the supplies are too low, in the wrong place and not suited to the forms of property that people now demand.

Commercial property

The central issue in the functioning of the commercial urban land market concerns the process of land assembly. Where public agencies are involved, the use of compulsory purchase powers ensures that land can be obtained and there is at least the prospect of some master-planning of the whole area. During the 1980s, the public agencies were quite important, especially in depressed urban areas, and land supply was therefore managed rather than simply left to market forces.

At other times and in prosperous areas where public agencies do not exist the process of land assembly has been left to market forces. In practice, some land-owners may wish to hoard land in anticipation of future increases in its value and in these circumstances the process of land purchase may be almost impossible.

An inevitable result of these problems is that developers prefer to purchase land on the outskirts of the urban area where large sites can be obtained from a single seller. The effect of this pressure on the urban fringe in Britain is often under-estimated. It also tends to further the process of urban decay.

Another issue associated with urban private sector land assembly is that many developments require the improvement, and possibly the development, of adjacent land. This is often impossible for property developers, the total site being much too large for their capabilities. In addition, urban developments may well require the creation of transport and other infrastructure, that the developer is unable to provide.

Overall, the commercial land assembly process continues to operate but in many urban areas the commercial sector is unable to function and urban dereliction tends to develop.

Clear problems are associated with the discretionary planning system, but the actors in the system have learned to cope with them and there is virtually no pressure to move to a legally binding plan. However, during the 1990s there has been a move to give the plan more significance in determining the acceptability of development applications, following the Planning and Compensation Act 1991.

10.2 Evaluation of the functioning of the market for urban property

Property in Britain is an asset that the owners and users expect to be able to exploit to their advantage. This statement holds true in both the owner-occupied housing market and the commercial sector. Treating property in this way implies that everyone associated with property needs to know the benefits and obligations associated with their interest in property. In the housing sector these are usually quite simple, but in the commercial market the complex landlord and tenant system, which generates leases of immense complexity, could create great difficulties for everyone.

One of the major attributes of the British system is the evolution of property professionals, particularly surveyors, able to give impartial advice to anyone about their legal responsibilities. In practice, all actors are aware of their obligations and the rewards they should get from their interest in property. Disputes are relatively rare, despite the complexity of the system.

The development process
It is the development process that generates most debate in Britain and that several governments have tried to "reform". The key elements of the system are:
– the planning system is flexible but also uncertain for developers;

- the planning system is subject to political control and the Secretary of State for the Environment has ultimate power to determine land usage;
- although user demands are important, the developer looks to the investment market for eventual purchase and therefore the state of the investment market is central to the development decision;
- private land assembly is the norm except for UDCs, etc. and the process is difficult and time consuming; furthermore each land-owner will try to exploit their land ownership to the full;
- land hoarding is not uncommon because owners have some incentive to wait for higher prices and few costs;
- public sector activity is particularly important in depressed regions and inner-urban areas;
- property is normally developed in anticipation of demand and therefore expectations are important;
- finance is always available for developments that are expected to be profitable.

These characteristics of the property development system have generated a process of development that is unique to Britain. It could not occur in another country where these characteristics did not exist. For example, the speculative application for development permission for a project that is not within the local plan occurs because flexibility in the system exists and because ultimately the Secretary of State may use his powers to approve the application. Furthermore, the apparent willingness of the current government to approve such applications actually encourages this behaviour.

The workings of the development process appear, to British eyes, to have certain weaknesses.

First, the system is highly cyclical. Whether this is a result of the cyclical nature of the British economy or whether it is a function of the property market is not clear. What is apparent is that in times of depression a large number of property developers become bankrupt and financiers of the development process lose a lot of money. Sites that have planning permission can be left derelict for many years, contributing to urban decay.

Secondly, the system is tied to the property investment market. If investors are not willing to buy, then development cannot take place. Furthermore, the investors determine the built form. Property must be flexible and capable of being re-let to other companies and not tied to the specific requirements of the user.

Thirdly, private land assembly in urban areas can be so difficult that developers are encouraged to look to "greenfield" sites. The use of compulsory purchase powers to assemble land for private development has never been widely accepted and still creates considerable debate when it occurs. One of the consequences of this is that the problem of inner-city decay cannot be solved by the private sector without major changes in the system. The UDCs and the EZs are attempts at new approaches but their impact is highly localized.

Fourthly, in depressed regions the lack of profitable development opportunities has resulted in major public action, especially in the industrial property sector. Such activity has ensured that the private sector will not return without public subsidy and therefore the market will not function freely. Even the current government has not been able to think of a way of returning to free market conditions. Although there are many short-term policy initiatives (UDCs, EZs, City Grants, National Garden Festivals, etc.), there is a lack of long-term strategic thinking, policy-making or investment.

Since 1990, there has been some acknowledgement of the rôle of strategic planning and environmental protection, but little tangible policy development so far. This is especially apparent in relation to the impact of the European single market and the opening of the Channel Tunnel in 1994.

Fifthly, users have been faced with accepting very demanding terms before they can obtain property. This has been caused by the dominance of the investment market in the post-1945 period. This may be coming to an end.

Sixthly, users often find that they cannot afford to obtain property in certain locations and therefore they are forced to locate in other, less attractive locations. This is perhaps inevitable in a market economy, but in Britain the local authorities have relatively few powers to ensure that certain types of land-use are maintained in their areas. For example, the change in the Use Classes Order has meant that land previously earmarked for industrial use can now be used for office space. Local authorities cannot stop the change, even where they have large numbers of unemployed industrial workers within their boundaries.

Lastly, the dominance of the private sector, especially during the 1980s, has meant that large-scale integrated urban developments, outside UDC and EZ areas, have been almost impossible. Few developers have the capabilities to plan and develop large areas and there is often a need for integration with transport and other public infrastructure. Developments have tended to occur where the existing infrastructure exists (e.g. along the line of the M25 motorway), even if this is all ready congested.

Overall, the system has certain characteristics that are clearly national specific. It does work. It does produce property to meet demands, but it is not without criticism, especially from the planning professionals, who find it increasingly difficult to influence or control.

The commercial property market

Many of the characteristics of the development process listed above are also relevant to the market in completed commercial buildings. The market is dominated by the investors, expectations of future conditions are of prime importance and finance is available for investments that hold out the prospect of good returns. Furthermore, the public sector is actively involved in certain regions/markets and this modifies market behaviour.

Property is an asset, a commercial commodity, much more than it is a means of production. Market behaviour is therefore dominated by the "desire" to obtain the best deal rather than the desire to meet the needs of the user. The market does not forget users however; after all, it is they who pay the rents and these and expectations of rental growth determine yields and eventually capital values.

The investment market does appear to have its own logic, leading to periods of rapid rises or falls in capital values. In this respect it is quite like the stock exchange, where the value of company shares can, for certain periods, be quite divorced from the performance of the company.

Within Britain, much wealth is now tied up in commercial property to the extent that few would wish to alter the system even if it could be shown to be unsuitable for the productive economy.

Housing

There is an enormous range of issues associated with the functioning of Britain housing sector that need to be considered. For example, the issues of homelessness, social housing, housing mobility and housing repairs all produce questions about the effectiveness of the British housing system. Some of these have been addressed elsewhere, so this section will concentrate on wider issues.

First, there is much evidence that for several decades there has been a lack of investment in the stock of housing in Britain. Although more houses are available, the growth has not kept up with the growth in households and the housing stock is falling into a poor state of repair. Major investment in the 1960s and 1970s in the public sector produced a stock of housing that is generally of poor quality and badly designed. Tenants generally do not like it and some of it has had a very short life. The problem of a lack of investment will eventually have to be faced.

Secondly, the growth in owner-occupation has been rapid and appears to coincide with most people's desires. It is now the dominant form of residential occupancy and it is possibly approaching saturation point. The last major boost was the 1980s "right to buy" for council tenants. Problems do exist, particularly the high costs of moving to prosperous regions, difficulties with sales and consequent impeding of mobility, encouragement to first-time buyers leading some to overreach themselves, leading to repossessions, and growing maintenance backlog. Nevertheless, there is little doubt that the public wish to occupy under this tenure. This is not surprising in Britain's inflationary environment.

What is becoming apparent in Britain is that it is not the size of the owner-occupied sector that matters but whether each sector is able to fulfil the functions for which it should cater. For example, a very small rental sector could be quite sufficient if it could provide the flexible type of housing that is required by many members of the community. For many people in Britain, the rental market is not an alternative because they cannot find a way of getting access to it. Either they are not a priority case for public housing or landlords will not rent because of rent control restrictions.

Thirdly, there is a growing proportion of the British population for whom the term "residualization" is appropriate. One of the real tests of housing policy in the 1990s will be whether it addresses the issue and finds solutions that are flexible enough to meet the complex housing needs that these groups create.

10.3 Conclusions of study

It is not intended to elaborate a long concluding section as this would repeat many of the points already made in the previous sections of the report. Instead, this section will offer some concluding remarks on the evaluation of the British property system in respect of the aims or desirable characteristics of urban land and property markets identified when work on the five-country project for the German government began.

> AIM 1: Within the public policy framework, the land and property market should secure that a sufficient supply of building land and property will be offered on each market sector.

At the outset it must be emphasized that the public policy framework during the 1980s emphasized that the use of land should be determined by market forces not public policy and that the planning system should restrict only development activity where there were clear planning reasons. It is a circular argument therefore to say that the land and property market secures a supply of land consistent with the public policy framework.

The flexibility of the British planning system tends to ensure that, where demand for land and property exists, development takes place unless restricted by the planning system. Plans themselves cannot stop development; only the willingness of the planning authorities to use the development control system to stop land being used for particular purposes, and the willingness of the Inspectorate and the SoS to uphold these decisions on appeal, will normally stop land supply matching demand.

There are some problems however. Land hoarding exists because some land-owners would rather wait for future increases in land values than sell in the current markets. Large-scale dereliction occurs in urban areas because the local authority does not have the finance with which to purchase the land and initiate large-scale development, while private land assembly is difficult and expensive and few developers would be able to master-plan large sites, especially in low-value areas.

Within the housing sector, the use of housing land availability (HLA) studies appeared to ensure that a sufficient supply of housing land existed for the construction of owner-occupied housing. The policy commitment to the provision of housing land is very strong, as it is seen as necessary for the continued growth of the owner-occupied sector.

> AIM 2: The land and property market should secure that all parts of the population have access to land-use and the market, having regard to geographical factors.

General issues of distribution have been important in Britain and one of the criticisms of the British political process and of industrial relations is that too much emphasis has been placed on the distribution of national output and too little on increasing it.

The issue of a broadly fair distribution of private property has not generally been significant. There has been a long-standing debate about the preferential tax treatment of owner-occupiers in the housing market because the results of the subsidy are that the highest paid have received the greatest benefit. There has also been some disquiet at the peak of the property cycle, where property owners appear to have made large capital gains even if these gains are not subsequently realized.

Geographical aspects of this have been particularly strongly felt because the peak of the property boom also represented the peak of the north/south differential in property prices. It is also clear that regional differences have inhibited mobility and therefore impeded both job hunting and recruitment at times of high unemployment.

The growth in owner-occupied housing has led to a substantial transformation of the ownership of wealth in Britain because the personal equity in people's housing commonly represents the bulk of their assets. Their other main asset is commonly their pension rights, which have been accumulated by investing with a pension fund. Much of these monies is invested in commercial property.

Access to other housing tenures is difficult and while the local authorities operate a system of priorities based on personal circumstances there is little doubt that they are unable to meet the needs of their communities. It is not the functioning of the land and property market that creates difficulties but rather that considerable additional resources are required.

In the commercial sector access, to land and property is generally via the market and, except in those locations where highly restrictive planning policies exist, the development industry has created an adequate supply to meet demand. Public intervention is possible within the system and has been used extensively in depressed regions to create advance factories that it is hoped will attract new firms to the locality.

Generally in the commercial sector firms accept that they have to pay the market price for the property they wish to obtain. In certain locations and at certain times this can involve very high prices, e.g. the London office market in 1989, but the government has never seriously considered restricting rents or dealing with the rationing of space that such a policy would imply.

Overall, the British attitude to property – that it is an asset that should be exploited – is more important than questions of equity.

> AIM 3: The land and property market should secure that the prices
> on the market are reasonable and appropriate, the volatility of the
> money market should be avoided.

The issue of volatility is highly pertinent to the British property market. The property cycles are both substantial and damaging to the development process and to the financial markets, and to the economy as a whole, because they encourage managerial resources into property management at the expense of productive activity.

What is not clear, however, is that the property cycles are created by the structures of the land and property market. Property is bound to be an unstable asset when persistent inflation exists in the economy and the level of economic activity itself is highly cyclical.

> AIM 4: The land and property market within its framework should
> secure that a careful and environmentally responsible conversion
> of land is promoted and achieved.

The UK system can secure this, but does not necessarily do so. It is entirely a question of political will on the part of local and central government.

The discretionary British system of planning control can be utilized to achieve responsible conversion of land. In practice, much depends on the policies set by local planning authorities, how consistently they are applied and enforced, and how far central government supports LPAs through appeal decisions and policy statements. Rigid codified rules are not applied. The use of green belts and density controls has had some beneficial effect in this area, but the dominance of market forces in development activity during the 1980s made these policies more difficult to maintain.

From 1990 to 1993 more social emphasis has been placed on environmental quality and it is quite possible that the development control process will respond to this change. What is lacking at the moment is the willingness to invest large sums of public or private money in improving environmental quality in urban areas, although the creation of the Urban Regeneration Agency in 1993 could have an impact.

> AIM 5: The framework of the land and property market should be such that private decisions concerning land and property can be influenced and regulated for the public weal.

The framework under which the public sector can influence and control the land and property market in Britain is quite clear. Planning controls can restrict development and compulsory purchase powers and public subsidies can promote certain forms of development. Certain major problems exist however.

To implement development control there needs to be a strong commitment to control by the LPAs, Inspectorate and SOS, especially in view of the discretionary powers. Therefore much depends in Britain on public policy decisions rather than the legal basis and procedures of the planning system. Again, the extent of influence exerted on private decisions for the public weal depends on political will. The framework is there.

Private developers have considerable difficulty in obtaining urban land and therefore much emphasis has been placed on developing land in areas outside the urban core, to the cost of other land-users, while vacant urban land persists.

Although planning authorities do have powers of compulsory purchase that could be used to obtain land for private development, the powers have not been extensively used. Part of the problem is that local authorities are unable to finance land purchase. This problem is exacerbated by the expectation of land-owners of high land prices and their perception of land as an investment. Another important issue is that the experience of compulsory land purchase and redevelopment during the 1950s and 1960s was that very poor public housing was created. Local authorities are not trusted, especially by the development industry, to use compulsory purchase in a way that would be compatible with commercial development.

The issue of who benefits from the increase in development value (betterment) associated with the giving of planning permission has been controversial in Britain for much of the post-1945 period. Both major political parties have generally been willing to tax increases in value, though not for the individual's first house, but the Labour Party has been willing to increase the effective tax rate for commercial development to 100%.

Currently taxation is limited to the standard rate of company tax and this represents the lowest rate of taxation possible.

The use of planning gain agreements is one method by which development value is taxed, but as a system it is highly inequitable because in some regions little benefit can be negotiated because of the low level of profitability of development. It can, at its worst, lead to planning permission for developments that should not receive approval on their planning merits.

> AIM 6: The land and property market and its trends should be visible and there should be a clear legal framework for the actors on the market.

Generally the British property system is clear and visible and the legal framework is well established. The availability of a range of professional advisers ensures that all those who participate in the process can obtain the information that they need and are therefore aware of their rights and responsibilities. This occurs despite the absence of an effective public information system and without a cadastre.

A major problem exists, however, in that the discretionary system of development control creates some uncertainty in the development process. Land-owners can never be sure of the response of the planning authority to their application for development approval. Flexibility is obtained at the expense of uncertainty.

APPENDIX

Policy guidance

The Department of the Environment issues Planning Policy Guidance notes and circulars on a continuous basis. The following were published or in preparation in 1993.

Planning Policy Guidance Notes

PPG1 *General Policy and Principles* Outlines the planning framework, purpose of planning system with general statement of planning policy

PPG2 *Green Belts* Explains purpose of the green belts and the need for general presumption against inappropriate development in the green belt

PPG3 *Land for Housing* Policy on provision of land for housing and role of planning system in meeting demand

PPG4 *Industrial and Commercial Development and Small Firms* Emphasizes importance of positive and prompt approach towards activities contributing to national and local economic activity

PPG5 *Simplified Planning Zones* Explanation of scheme

PPG6 *Major Retail Development* Policy on large new retail development

PPG7 *Rural Enterprise and Development* Advice on non-agricultural development in the countryside, including new uses of buildings

PPG8 *Telecommunications* Advice on planning aspects of telecommunications development

PPG9 *Regional Guidance for the South East*

PPG10 *Strategic Guidance for the West Midlands*

PPG11 *Strategic Guidance for Merseyside*

PPG12 *Local Plans* Underlines importance to be attached to up-to-date local plans as basis for planning control

PPG13 *Highway Considerations in Development control*

PPG14 *Development on Unstable Land*

PPG15 *Regional Planning Guidance, Structure Plans and the Content of Development Plans* (now cancelled)

PPG16 *Archaeology and Planning*

PPG17 *Sport and Recreation*

PPG18 *Enforcing Planning Control*

PPG19 *Outdoor Advertisement Control*

PPG20 *Coastal Planning*

PPG21 *Tourism*

PPG22 *Renewable Energy*

APPENDIX

Proposed PPGs

Noise
Nature Conservation
Planning, Pollution Control and Waste Management
Listed Buildings and Conservation Areas

Minerals Policy Guidance Notes

MPG1 General Considerations and the Development Plan System
MPG2 Applications, Permissions and Conditions
MPG3 Opencast Coal Mining
MPG4 The Review of Mineral Working Sites
MPG5 Minerals Planning and the General Development Order
MPG6 Guidelines for Aggregates Provision in England and Wales
MPG7 cancelled
MPG8 Planning and Compensation Act 1991: Interim Development Order Permissions, Statutory Provisions
MPG9 Planning and Compensation Act 1991: Interim Development Order Permissions, Conditions
MPG10 Provision of Raw Material for the Cement Industry

Development Control Policy Notes

DCPN1 General Principles
DCPN2 Development in Residential Areas
DCPN3 Industrial and Commercial Development
DCPN4 Development in Rural Areas
DCPN5 Development in Town Centres
DCPN6 Road Safety and Traffic Requirements
DCPN7 Preservation of Historic Buildings and Areas
DCPN8 Caravan Sites
DCPN9 Petrol Filling Stations and Motels
DCPN10 Design
DCPN11 Amusement Centres
DCPN12 Hotels and Motels
DCPN13 Large New Stores (Retail)
DCPN14 Warehouses
DCPN15 Hostels and Homes
DCPN16 Access for the Disabled.

Circulars are issue on a continuous basis by the appropriate government department and form the most immediate guide to current policy. Their contents are often restated in updated PPGs, MPGs and DCPNs.

The General Development Order, Permitted Development: Summary

    I Development within the curtilage of a dwelling house (subject to qualifications relating to the height, set back, plot coverage and cubic volume)
   II Gates, fences etc., external painting
  III Changes of use within use classes defined in Use Classes Order
  IV Temporary buildings and uses
   V Use by members of specified recreational organisations
  VI Agricultural buildings, works and uses
 VII Forestry buildings and uses
VIII Development on land used for industrial purposes and uses (subject to qualifications as for I)
  IX Repairs to private roads
   X Repairs to services (sewers etc.)
  XI Replacement of war-damaged buildings
 XII Development under private Acts or Orders
XIII Development by local authorities and statutory undertakers, such as
–XXIII repairs and improvements to highways

The Use Classes Order: Summary

A1 Shops
A2 Financial and professional services
A3 Food and drink
B1 Business including offices, research and development and light industrial
B2 General industrial building
B3 Special industrial: alkalis etc.
B4 Special industrial; smelting etc.
B5 Special Industrial: bricks etc.
B6 Special industrial: chemicals etc.
B7 Special industrial: animal products etc.
B8 Storage and distribution
C1 Hotels and hostels
C2 Residential institutions
C3 Dwelling-houses
D1 Non-residential institutions
D2 Assembly and leisure

Census

The 1991 Census requested the following information:

APPENDIX

Personal topics

1 Name
2 Address at time of census
3 Usual address
4 Sex
5 Date of birth
6 Marital status
7 Relationship in household
8 Country of birth
9 Scottish Gaelic language (Scotland only)
10 Welsh language (Wales only)
11 Ethnic group[a]
12 Address one year ago
13 Term-time address of students[a]
14 Economic position/employment status
15 Occupation
16 Industry of occupation
17 Hours worked weekly[a]
18 Address of work place
19 Means of travel to work place
20 Higher education qualifications
21 Long-term illness[a]
22 Improved resident population base[ab]

Household topics

23 Dwellings classification and count[ab]
24 Type of accommodation
25 Tenure
26 Number of rooms
27 Bathroom and WC
28 Central heating[a]
29 Number of cars and vans
30 Lowest floor level of accommodation (Scotland only)

[a] Not included in 1981 census
[b] New feature not requiring a specific question

BIBLIOGRAPHY

Aldridge, M. 1979. *The British new towns*. London: Routledge.

Ashworth, W. 1968. *The Genesis of modern British town planning*. London: Routledge.

Balchin, P. N. 1989. *Housing policy. An introduction*, 2nd edn. London: Routledge.

Barnes, J. 1990. Urban Development Corporation – the lessons from London's Docklands. In *Radical planning initiatives*, J. Montgomery & A. Thornley (eds), 59–94. Aldershot: Gower.

Barrett, S. & P. Healey (eds) 1985. *Land policy: problems and alternatives*. Aldershot: Gower.

Baum, A. & N. Crosby 1988. *Property investment appraisal*. London: Routledge.

Boleat, M. 1986. *The building society industry*, 2nd edn. London: Allen & Unwin.

Boleat, M. & A. Coles 1987. *The mortgage market*. London: Allen & Unwin.

Brazier, R. 1988. *Constitutional practice*. Oxford: Oxford University Press.

Brindley, T., Y. Rydin, G. Stoker 1989. *Remaking planning*. London: Unwin Hyman.

Brownhill, S. 1990. *Developing London's docklands*. London: Paul Chapman.

Bruton, M. & D. Nicholson 1987. *Local planning in practice*. London: Hutchinson.

Burn, E. H. 1988. *Cheshire & Burn's modern law of real property*, 14th edn. London: Butterworth.

Butterworths 1989. *Butterworths property law handbook*, 2nd edn (ed. E. H. Scamell). London: Butterworth.

Butterworths 1992. *Butterworths planning law handbook*, 3rd edn (ed. B. Greenwood). London: Butterworth.

Cadman, D. & L. Austin-Crowe 1991. *Property development*, 3rd edn, ed. R. Topping & M. Avis. London: E. & F. N. Spon.

Carnwath, R., G. Hart, A. Williams 1990. *Blundell & Dobry's planning appeals & inquiries*, 4th edn. London: Sweet and Maxwell.

Champion, A. G. & A. R. Townsend 1990. *Contemporary Britain. A geographical perspective*. London: Edward Arnold.

Cherry, G. 1974. *The evolution of British town planning*. London: Leonard Hill.

Cherry, G. 1982. *The politics of town planning*. Harlow: Longman.

Cullingworth, J. B. 1988. *Town and country planning in Britain*, 10th edn. London: Unwin Hyman.

Darlow, C. (ed.) 1983. *Valuation and investment appraisal*. London: Estates

Darlow, C. (ed.) 1983. *Valuation and investment appraisal*. London: Estates Gazette.

Davies, H. W. E., D. Edwards, A. J. Hooper, J. V. Punter 1989. *Planning control in Western Europe*. London: HMSO.

Denyer-Green, B. 1989. *Compulsory purchase and compensation*, 3rd edn. London: Estates Gazette.

Distributive Trades EDC 1988. *The future of the high street*. London: HMSO.

DoE 1989. *Environmental assessment: a guide to procedures*. London: HMSO.

DoE 1990. *This common inheritance: Britain's environmental strategy*. London: HMSO, Cmnd 1200.

DoE 1992. *Local plan preparation: best practice guidance*. HMSO, London.

Donnison, D.& D. Maclennan 1991. *The housing service of the future*. Harlow: Institute of Housing/ Longman.

Elcock, H. 1986. *Local government politicians, professionals and the public in local government*. London: Methuen.

Elson, M. J. 1986. *Green belts: conflict mediation in the urban fringe*. London: Heinemann.

Ennis, F, P. Healey, M. Purdue 1993. *Frameworks for negotiating development*. Newcastle upon Tyne: Department of Town and Country Planning, University of Newcastle upon Tyne.

Fothergill, S., S. Monk, M. Perry 1987. *Property and industrial develop ment*. London: Hutchinson.

Fraser, W. D. 1984. *Principles of property investment and pricing*. London: Macmillan.

Glasson, J. 1992. *An introduction to regional planning*, 2nd edn. London: UCL Press Ltd.

Glasson, J., R. Therivel, A. Chadwick 1994. *Environmental impact assessment*. London: UCL Press.

Grant, M. 1982. *Urban planning law* (with 1990 supplement). Sweet & Maxwell, London.

Gravells, N. P. 1992. *Sweet & Maxwell's property statutes*, 6th edn. London: Sweet & Maxwell.

Greed, C. 1993. *Introducing town planning*. Harlow: Longmans.

Grimley J.R. Eve 1992. *The use of planning agreements*. London: HMSO.

Hall, P. 1989. *Urban and regional planning*. London: Unwin Hyman.

Hall, P., H. Gracey, R. Drewett, R. Thomas 1977. *The containment of urban England*. London: Allen & Unwin.

Hallett, G. & R. H. Williams 1988 West Germany. In G. Hallett (ed.) *Land and housing policies in Europe and the USA: a comparative analysis*, 17–48. London: Routledge.

Hamnett, C. & B. Randolph 1988. *Cities, housing and profits*. London: Hutchinson.

Healey, P. 1983. *Local plans in British land use planning*. Oxford:

Pergamon.

Healey, P. & R. Nabarro 1990. *Land and property development in a changing context*. Aldershot: Gower.

Healey, P., P. McNamara, M. Elson, A. Doak 1988. *Land use and the mediation of urban change*. Cambridge: Cambridge University Press.

Healey, P., S. Davoudi, M. O'Toole, S. Tavsanoglu, D. Usher (eds) 1992. *Rebuilding the city: property-led urban regeneration in Britain in the 1980s*. E. & F. N. Spon, London.

Heap, D. 1991. *An outline of planning law*, 10th edn. London: Sweet & Maxwell.

Hertfordshire Structure Plan 1986. Hertford: Hertfordshire County Council.

Hertsmere 1990. *Official guide*. Gloucester: British Publishing Co Ltd.

HMSO 1992. *The effects of major out of town retail development. Literature review for the Department of the Environment*. London: HMSO.

Holliday, I., G. Marcou, R. Vickerman 1991. *The Channel Tunnel: public policy, regional development and European integration*. London: Belhaven.

Imrie, R. & H. Thomas 1992. *Urban policy and city regeneration*. London: Paul Chapman.

Jordan, A. G. & J. J. Richardson 1987. *British politics and the policy process*. London: Unwin Hyman.

Jowell, J. & D. Oliver D (eds) *The changing constitution*, 2nd edn. Oxford: Clarendon.

Karn, V., J. Kemeny, P. Williams 1985. *Home ownership in the inner city. Salvation or despair?* Aldershot: Gower.

Laffin, M. 1986. *Professionalism and policy: the role of the professions in the central–local government relationship*. Aldershot: Gower.

Lawless, P. 1979. *Urban deprivation and government initiative*. London: Faber & Faber.

Lawless, P. 1989. *Britain's inner cities*. London: Paul Chapman.

Little, A. J. 1992. *Planning controls and their enforcement*, 6th edn. Crayford: Shaw & Sons.

McKenzie, J. A. & M. Phillip 1989. *Practical approach to land law*. London: Blackstone.

Marriot, O. 1989. *The property boom*. London: Abingdon.

Merrett, S. 1982. *Owner-occupation in Britain*. London: Routledge.

Morgan, P. & A. Walker 1988. *Retail development*. London: Estates Gazette.

Municipal Yearbook 1992. London: Municipal Yearbook Ltd.

Nadin, V. & S. Jones 1990. A Profile of the Profession. *The Planner* **76**(3), 13–24.

Purdue, M., E. Young, J. Rowan-Robinson 1989. *Planning law and procedure*. London: Butterworth.

Richards, P. G. 1988. *Mackintosh's the government and politics of Britain*, 7th edn. London: Unwin Hyman.

Robinson, J. F. F. 1988. *Post-industrial Tyneside*. Newcastle upon Tyne:

Newcastle City Libraries.

Rodriguez-Bachiller, A. 1988. *Town planning education: an international survey*. Aldershot: Gower.

RTPI 1991. *The education of planners*. London: Policy statement of the Royal Town Planning Institute.

Rydin, Y. 1986. *Housing land policy*. Aldershot: Gower

Rydin, Y. 1993. *The British planning system. An introduction*. London: Macmillan

Skeffington, A. 1969. *People and planning*. London: HMSO.

Telling, J. 1990. *Planning law and procedure*. London: Butterworth.

Thornley, A. 1990. *Urban planning under Thatcherism*. London: Routledge.

Townsend, A. R. 1993. *Uneven regional change in Britain*. Cambridge: Cambridge University Press.

Ward, M. 1990. *The Local Government and Housing Act 1989*. Coventry: Institute of Housing.

Whitaker 1992. *Whitaker's Almanack*, 124th edn. London: Whitaker.

Wood, B. & R. H. Williams 1992. *Industrial property markets in Western Europe*. London: E. & F. N. Spon.

INDEX

empirica

Gesellschaft für Struktur- und
Stadtforschung mbH
Kaiserstraße 29
53113 Bonn
Telefon (0228) 91 48 9-0
Telefax (0228) 21 74 10

Immobilienmärkte in Großbritannien

Verfasser:

Ulrich Pfeiffer
Volker Gillhaus
Boris Braun

9011
Bonn, Oktober 1991/Juni 1993

Contents

Vorbemerkung

Der nachfolgende Text wurde im Sommer 1991 abgeschlossen. Es bestand keine Möglichkeit, die Daten über Bautätigkeit und Investitionsvolumen zu aktualisieren. Deshalb wurde der entsprechende Text gekürzt und nur in wesentlichen Ergebnissen zusammengefaßt dargestellt. Auf diese Weise wurde sichergestellt, daß sich die Aussagen stärker auf strukturelle Charakterisierungen des Marktes konzentrieren, die sich im Zeitablauf nur sehr allmählich ändern.

TEIL A
Rückblick und Analyse der Situation

1
Zur volkswirtschaftichen Relevanz und zu einigen Besonderheiten von Immobilienmärkten

Die ökonomische Bedeutung von Immobilienmärkten

Das Immobilienvermögen stellt mit Abstand den größten Teil des Sachvermögens in einer Volkswirtschaft dar. Analog dazu gilt gleichzeitig, daß Bauinvestitionen einen beachtlichen Teil der jährlichen Bruttoanlageinvestitionen ausmachen. Der Vermögensanteil der Immobilien ist jedoch höher als ihr Anteil an den laufenden Bruttoinvestitionen, weil Gebäude wegen ihrer langen Lebensdauer ein höheres Gewicht in der Vermögensbilanz der Volkswirtschaft erhalten als Investitionen vergleichbarer Höhe mit geringer Lebensdauer.

Immobilienmärkte unterscheiden sich von anderen Gütermärkten durch typische Merkmale, die auch zu anderen Funktionsweisen und Abläufen der Märkte führen. Als zentraler Grund ist zu nennen, daß Grundstücke anderen Wettbewerbsbedingungen unterliegen als etwa Pkws. Die Produzenten von Pkws stehen auf ihren Absatzmärkten Monat für Monat mit anderen Konkurrenten im Wettbewerb. Sobald ihre Produkte relativ an Attraktivität verlieren, äußert sich dies in Marktanteilsverschiebungen. Auf solche Marktanteilsverschiebungen müssen die Unternehmen in der der Regel rasch reagieren. Entweder müssen sie ihre Produkte attraktiver gestalten oder die Preis-Kostensituation muß relativ günstiger werden. Solche unmittelbaren Wettbewerbswirkungen bestehen am Markt für Grundstücksnutzungen nur z.T. und noch weniger am Markt für Immobilien. Die Ursache liegt in der langen Lebensdauer der Objekte bzw. in der Tatsache daß Boden überhaupt nicht abgeschrieben werden muß und Eigentümer über ein ewiges Nutzungsrecht verfügen. In der Regel werden Vermieter durch die hohen Finanzierungskosten gezwungen in den Anfangsjahren der Bewirtschaftung den Markttendenzen sehr zeitnah zu folgen, weil sie nur auf diese Weise eine Rentabilität oder Liquidität des Investitionsprojekts sichern können. Auf Dauer sinken vor allem die Finanzierungskosten relativ zu den Mieten ab. Es entstehen Überschüsse und Spielräume. Das Nichtausschöpfen dieser Spielräume führt jedoch nicht, wie auf anderen Märkten, relativ rasch zu Verlusten. Die Eigentümer verzichten allerdings auf Renditen. Aufgrund ihres Eigentumsrechts werden sie durch den Wettbewerb nicht gezwungen, die jeweiligen Standortpotentiale maximal auszuschöpfen. Sie können, ohne dafür im Wettbewerb direkt bestraft zu werden, unterhalb der Marktmieten Verträge beibehalten. Daß solche Vermietung häufig vorkommen, zeigt allein die Tatsache, daß

Tabelle 1: Reproduzierbares Anlagevermögen in Westdeutschland[1] in Preisen von 1980 (in Mrd. DM).

	1960	in %	1970	in %	1980	in %	1988	in %
Primärer Sektor	298,8	11,7	445,5	9,9	597,9	8,8	700,1	8,3
Sekundärer Sektor	357,1	13,9	718,0	16,0	987,9	14,5	1.080,9	12,8
davon								
Verabeitendes Gewerbe	337,3	13,2	662,1	14,7	918,4	13,5	1.017,2	12,0
Tertiärer Sektor	1.905,8	74,4	3.775,8	83,9	5.891,7	86,7	7.525,9	88,8
davon								
Handel	82,4	3,2	168,4	3,7	246,9	3,6	290,4	3,4
Kreditinstitute u. Versicherungen	29,6	1,2	58,9	1,3	104,1	1,5	146,4	1,7
Sonstige Dienstleistungen	74,6	2,9	182,1	4,0	420,4	6,2	742,2	8,8
Alle Wirtschaftszweige	2.561,8	100,0	4.499,2	100,0	6.792,3	100,0	8.467,3	100,0

Quelle: Statistisches Bundesamt

1, Das reproduzierbare Anlagevermögen umfaßt auch Maschinen und Geräte. Gerade in Bereichen wie Handel, Kreditinstitute oder sonstige Dienstleistungen überwiegt das Immobilienvermögen. Das reale Wachstum in dieses Bereichen ist gleichzeitig Indikator für das Wachstum von Immobilienvermögen.

die Mietstruktur am Markt für Gebäudemieten eine sehr große Bandbreite aufweist. Dies ist einmal Folge der geringen Markttransparenz, daneben aber Folge der üblichen Trägheiten bei der Anpassung der Preise an veränderte Knappheiten. Aus diesen Preisanpassungsträgheiten entstehen Trägheiten in der Strukturanpassung bzw. in der Nutzungskonkurrenz. Hinzu kommt, daß in der politischen Öffentlichkeit Verdrängungen generell als etwas Negatives angesehen werden. Tatsächlich machen Verdrängungen am Markt für gewerbliche Mieträume nur deutlich, daß die neuen Konkurrenten die vorhandenen Standortpotentiale im Rahmen der jeweiligen planerischen Vorgaben besser ausschöpfen.

In der politischen Diskussion werden solhe Anpassungsträgheiten oft als erwünscht angesehen. Sie vermindern für die Nutzer die Härten des Wettbewerbs. Diese Position ist ohne Zweifel für den Wohnungssektor richtig, weil Mobilität hier jeweils mit persönlichen Belastungen verbunden ist. Im gewerblichen Bereich, insbesondere im Einzelhandel, führen Anpassungsträgheiten jedoch zu unterausgeschöpften Standortpotentialen und damit zur Nicht-Ausschöpfung von Infrastrukturleistungen und sonstigen komplementären Investitionen. Auch diese Argumentation wird nicht allgemein akzeptiert werden. So kann man feststellen, daß etwa in den Innenstädten in den Fußgängerzonen der Wettbewerb der Einzelhandelsunternehmen dazu führt, daß weniger leistungsstarke Geschäfte in Randbereiche abgedrängt werden. Daraus entsteht oft eine Einengung der Bandbreite der Angebote und eine Verarmung der Erlebnisvielfalt von Innenstädten. Dies wird häufig mit Recht beklagt. Es wäre jedoch keine Lösung durch Abbau von Wettbewerb, d.h. durch Mieterschutzgesetzgebung historisch-politisch erwünschte Nutzungen und Nutzungsstrukturen zu sichern. Die Kritik an solchen Entwicklungen macht inkonsistente Planungskonzepte sichtbar. Die Stadtplanung wünscht auf der einen Seite eine hohe Zugänglichkeit der Innenstädte, etwa durch innenstadtorientierte Nahverkehrs-systeme. Sie lehnt jedoch die Folgen der erhöhten Erreichbarkeit, etwa im Einzelhandels- bzw. im Bürosektor zum Teil ab. Diese Position ist dann gerechtfertigt, wenn sich zeigen läßt, daß positive externe Effekte bestimmter Nutzungen (Traditionsgeschäfte, die Touristen anlocken, bzw. deren Auslagen von den Besuchern als Bereicherung empfunden werden) durch Wettbewerb verdrängt werden. Dennoch wäre es unsystematisch und nicht sehr effizient, solche erwünschten Nutzungen durch Mieterschutz und Anpassungsträgheiten zu konservieren. Durch einen allgemeinen Mieterschutz und allgemeine Anpassungsträgheiten werden auch Nutzungen ohne positive externe Effekte begünstigt. Eine rationale Steuerung müßte hier gezielte Standortsubventionen für Nutzer mit positiven externen Effekten entwickeln.

Mit diesem Hinweis soll gleichzeitig verdeutlicht werden, daß es rational ist, Rahmenbedingungen zu setzen, die einen scharfen Leistungswettbewerb hervorrufen. Dort, wo die Ergebnisse dieses Leistungswettbewerbs mit dem Hinweis auf externe Effekte oder gemeinnützige Folgewirkungen nicht erwünscht sind, müssen gezielte Standortsubventionen gewährt werden. Dies kann z.B. in der Form geschehen, daß die Städte selbst an bestimmten Standorten durch Globalanmietungen Nischen reduzierten Wettbewerbs sichern und die erwünschten Nutzungen in Untermietverträgen mit tragbaren Mieten erhalten. Die Finanzierungsmittel für solche politisch erwünschten Nutzungen könnten z.B. im Rahmen eines City Management durch Sonderabgaben von den übrigen Nutzern erhoben werden, die keine positiven externen Effekte erzeugen, bzw. sogar negative Ausstrahlungen auf die Attraktivität der Stadtbereiche haben.

Unabhängig von Anpassungsträgheiten, die aus den besonderen Wettbewerbsbedingungen an Immobilienmärkten entstehen, kommen weitere Faktoren hinzu. So haben in der Bundesrepublik «Amateurinvestoren» bzw. private Eigentümer hohe Marktanteile am Markt für Bürogebäude. Die Kombination aus privatem Familienvermögen und schwachem Leistungswettbewerb verstärkt die Tendenz zu Anpassungsträgheiten bei der Veränderung von Lagequalitäten bzw. von konjunkturellen Knappheiten.

Verstärkung der Trägheitsphänomene

Die Trägheitsphänomene werden am Markt für Immobilien und Immobiliennutzungen durch steuerliche Regelungen verstärkt. Dies gilt insbesondere für den Eigentümermarkt. Firmen, die gleichzeitig Eigentümer ihrer Fabrikgebäude oder sonst genutzten Gebäude sind, halten an historischen Standorten lange fest, weil sie Grundvermögen zum Niedrigswertprinzip bilanzieren und die Verkehrswertsteigerungen lediglich zu wachsenden stillen Reserven führen. Untergenutzte Grundstücke führen auch nicht zu hohen Grundsteuerbelastungen. Im Ergebnis kommt es zu Verlagerungen fas immer erst dann, wenn in den bestehenden Grundstücksgrenzen eine weitere Expansion nicht möglich ist oder wenn sich die Verkehrssituation so sehr verschlechtert, daß eine Verlagerung opportun wird. Standortwechsel werden deshalb von Eigennutzern nur in sehr langen Fristen vorgenommen. In der Umkehrung kann man beobachten, daß in fast allen älteren Gewerbegebieten erheblich Reserveflächen bestehen. Die ursprünglichen Expansionsabsichten wurden nicht verwirklicht. Die jetzt bestehenden Grenzen und Eigentumsrechte und der fehlende Wettbewerb liefern keinen Anreiz zur besseren Ausnutzung von Standortpotentialen. Unter diesen Bedingungen kommt es dann zu längeren Flächenhortungen. Dabei spielt eine Rolle, daß diese stillen Reserven ohne Belastungen gehalten werden können.

Volkswirtschaftlich führt der schwache Wettbewerbsdruck dazu, daß strukturelle Veränderungen der Nutzung von knappen Immobilien sich nur verzögert durchsetzen. Dies gilt am Markt für Nutzungsrechte (Mietermarkt) weniger als am Eigennutzermarkt. Die Anreize für das Horten von Grundstücken und das Halten von stillen Reserven sind sehr hoch. Die Anreize für Standortpotentiale möglichst ohne Zeitverzögerung auszuschöpfen, sind gering. Es liegt auf der Hand, daß es sehr schwer ist, eine optimale Anpassungsgeschwindigkeit an veränderte Lagebedingungen zu definieren. So spielt der Restwert von Anlagen, die mit dem Boden verbunden sind, sicherlich eine erhebliche Rolle bei der Bestimmung der künftigen Nutzung. In jedem Fall sind die gegenwärtigen steuerlichen Rahmenbedingungen jedoch wenig geeignet, um Anpassungen an veränderte Lagefaktoren zu beschleunigen.

Die Internationalisierung der Immobilienmärkte zwingt zur Anpassung der Rahmen-bedingungen.

In den letzten Jahren haben sich die Immobilienmärkte mehr und mehr internationalisiert. In der Bundesrepublik wurden vor allem Frankfurt, München, Hamburg und jetzt Berlin zu Zielorten für internationale Immobilieninvestoren. In Berlin kann man davon ausgehen, daß ein hoher Teil der Büroinvestitionen von ausländischen Firmen, d.h. von professionellen Anlegern errichtet werden. Diese Internationalisierung der Märkte führt auch zu einer Internationalisierung der entsprechenden Dienstleistungen, der Planungsteams und der Bewirtschaftung. Komplexe Bauvorhaben, insbesondere Hochhäuser, werden von «Baunomaden» geplant und gesteuert. Planungen dieser Art setzen leistungsfähige, eingespielte Teams voraus, die es in dieser Größenordnung in der Bundesrepublik nicht gibt. Deshalb stellt man fest, daß Spezialisten, die einige Jahre in Canary Wharf gearbeitet haben, später z.T. bei den gleichen oder bei anderen Firmen in Frankfurt oder auch in Paris beschäftigt sind. Eine Konsequenz dieser Internationalisierung besteht darin, daß große Baukomplexe in ihrer Technologie bis hin zur Grundrißgestaltung ebenfalls internationalisiert werden. Lokale Bautraditionen werden zurückgedrängt. Dies äußert sich gegenwärtig in Berlin in der Diskussion um die Erhaltung des Berliner Baublocks und die planerischen Vorgaben, z.B. um traditionelle Traufhöhen zu erhalten. Die internationalen Investoren haben wenig Neigung, auf solche Vorgaben einzugehen, weil dadurch «eingefahrene» Bautechniken und Bauformen nicht nahtlos angewandt werden können. Von anderen wird darauf verwiesen, daß heute ein hohes Maß von Flexibilisierung in den Bauformen erreicht ist, mit der Folge, daß es leicht möglich sei, sich an regionale Bautraditionen anzupassen.

In den Bauordnungen und sonstigen technischen Vorgaben zur Gestaltung der Gebäude entsteht eine Tendenz zur Vereinheitlichung. Auch die Dienstleistungen um die Immobilien bzw. am Immobilienmarkt vereinheitlichen sich. In der Bundesrepublik ist dies durch die Expansion englischer Maklerfirmen sichtbar geworden (Beispiel: Jones Lang Wootton verfügt inzwischen über Repräsentanzen in Frankfurt, Hamburg, Düsseldorf, Berlin und München; Healey und Baker über Repräsentanzen in Düsseldorf, Frankfurt und Berlin).

Es liegt auf der Hand, daß anders als bei mobilen Gütern der Zwang zur Vereinheitlichung von Rahmenbedingunen weit geringer ist. Man kann den Standpunkt vertreten, daß die jeweiligen ausländischen Investoren sich bei Immobilieninvesitionen und ihrer Bewirtschaftung an die lokalen Verhältnisse anzupassen haben. Dieser Position sollte mit Recht überall dort Geltung verschafft werden, wo es darum geht, lokale Besonderheiten, insbesondere aber auch von der Öffentlichkeit als positiv empfundene Ergebnisse der Bau- und Immobilienmärkte zu erhalten. Die Position läßt sich auf Dauer dort kaum aufrechterhalten, wo wettbewerbsvermindernde Regelungen bestehen, wo Produktivitätssteigerungen oder Effizienzsteigerungen bei der Ausschöpfung von Nutzungspotentialen nicht voll realisiert werden. Vereinheitlichungen von Rahmenbedingungen dürften es jedoch nicht unmöglich machen, daß lokale Besonderheiten des Stadtgrundrisses, der Bauformen und der Bautraditionen weiterhin bewahrt werden. Die Attraktivität europäischer Städte liegt in diesen Besonderheiten. Eine Internationalisierung der Städte durch Internationalisierung der Rahmenbedingungen durch Berufsordnungen, Bauordnungen und anderer normativer Vorgaben kann nicht im Interesse der nationalen Politik liegen. Dadurch ist deutlich gemacht wie ambivalent die Ergebnisse von vereinheitlichten Regeln sein können.

Die Entwicklung der Immobilienmärkte in Großbritannien – quantitative Analyse[1]

Der folgende Text enthält die gekürzte Fassung eines längeren Analysekapitels. Die Aussagen konzentrieren sich auf einige wenige herausragende Merkmale der Entwicklung des Marktes.

Starke Fluktuationen

Ein hervorstechendes Merkmal des englischen Immobilienmarktes war in der Vergangenhein seine starken Fluktuationen. Stärker noch als die Gesamtwirtschaft schwanken die Preise, die Renditen aber auch die Bautätigkeit und die Umsatzhäufigkeit am Markt für Immobilien. Dabei spiegeln die Immobilienzyklen die allgemeine Wirtschaftsentwicklung nur zum Teil wider, obwohl eindeutige Zusammenhänge bestehen.

Großbritannien hat zu Beginn der 80er Jahre eine weit ausgeprägtere Stabilisierungs- und Strukturanpassungskrise erlebt als die Bundesrepublik. Das Bruttosozialprodukt ist über mehrere Jahre erheblich gesunken. Seit Mitte der 80er Jahre kam es in Großbritannien zu einem extremen Bauboom. Er wird begleitet von einem hohen Anstieg der realen Mieten. Die kräftigen Mietsteigerungen seit 1986 beendeten eine Periode langfristiger sinkender Realmieten am Immobilienmarkt. Dieser Abstieg begann mit der Ölkrise und hat sich erst mit dem Aufstieg in den 80er Jahren deutlich verändert. Am ausgeprägtesten war dieser Bauboom im Londoner Büromarkt. In den Jahren 1981/82 vor dem Ausbruch der Krise wurden allein in der City von London etwa 350.000qm Büroflächen fertiggestellt. Diese Fertigstellungen sanken auf ein Niveau zwischen 100.000 und 140.000qm jährlich ab, um dann später allein in der City auf Größenordnungen von 500.000qm anzusteigen. Wie wir inzwischen wissen, kam es nach 1989/90 zu einem aprubten Ende dieses Baubooms und Leerstandsraten, wie sie in der Bundesrepublik nie erlebt wurden.

Extreme Schwankungen dieser Art sind nicht nur Folgen reiner Marktfaktoren. In Großbritannien kamen als exogene Ereignisse hinzu der sogenannte «Big Bang», d.h. die Deregulierungen der Aktien- und Finanzmärkte im Jahre 1986. Hier wurde ein erheblicher Nachfrageschub nach gut ausgestatteten Büros ausgelöst.

Es kam zu regionalen Konkurrenzen. Die neuen Standorte, z.B. in den Docklands, drohten für die City, einen relativen Bedeutungsverlust hervorzurufen. Dies führte dazu, daß die City mit einer Ausweitung der Baurechte antwortete. Diese regionale Konkurrenz hat Überinvestitionen begünstig.

Neue Kreditgeber waren bereit, unter Bedingungen scharfen Wettbewerbs höhere Risiken als in der Vergangenheit zu tragen.

Am Investitionsboom haben sich auch im erheblichen Umfang neue Investoren beteiligt. Während die traditionellen Fonds ihre Investitionen nur mäßig steigerten, haben vor allem ausländische «Newcomer» den Boom «angeheizt».

Investoren und Kreditgeber zusammen haben die Aufnahmefähigkeit des Marktes in einem extremen Ausmaß unterschätzt. abei ist allerdings zu berücksichtigen, daß die Stabilisierungskrise, die mit Ende der 80er Jahre einsetzte, schwer vorherzusehen war.

Die starken zyklischen Schwankungen am englischen Immobilienmarkt sind auch das Ergebnis eines ausgeprägten «Anpassungsverhalten» der öffentlichen Hand. In der Bundesrepublik kann man immer wieder beobachten, daß die Kommunen eine stark steigende zyklische Nachfrage nur zögerlich in Baurechte umsetzen. D.h. das Planungsverhalten der Kommunen führt zu einer gewissen «Streckung» von Planungen. Im nachhinein läßt sich feststellen, daß aus dem Zusammenspiel zwischen privaten Investoren und der Bereitstellung von Infrastruktur und Baugenehmigungen eine Dämpfung von Zyklen herbeigeführt wurde. Zyklusverstärkende Verhaltensweisen der öffentlichen Genehmigungsbehörden, wie sie in London zu beobachten waren, sind in der Bundesrepublik kaum aufgetreten. Eine Ausnahme bildet in jüngster Zeit das Planungsverhalten in Berlin. Hier verstärkt die öffentliche Hand durch die Vorbereitung zahlreiche Projekte die Durchführung von Wettbewerben und anderer Maßnahmen, die ohnehin an den überzogenen Erwartungen der Investoren. Ein solches Verhalten ist jedoch atypisch.

Die Marktstruktur
– Vergleich zur Bundesrepublik

Unterschiede auf der Anbieterseite

Starke Stellung professioneller Investoren in Großbritannien
Die Immobilieninvestoren in Großbritannien gehen weitgehend von anderen Vorausset-
zungen und Zielen aus, als in der Bundesrepublik. In Großbritannien wird der Markt
(Neubau und Bestand von Büro, Gewerbe und Industrieimmobilien) für vermietete
Objekte weitgehend von den institutionellen Anlegern, Versicherungen und Pension
Funds sowie von den Property Companies und einzelnen Großinvestoren aus dem
Ausland beherrscht. Von den rd. 90 Mrd. £ an vermieteten gewerblichen Immobilien-
investitionen 1990 gehören etwa 65 Mrd. £ den institutionellen Anlegern (Versich-
erungen, Pension Funds) und etwa 25 Mrd. £ den Property Companies. Dies entspricht
etwa 25 % des Wertpapiervolumens, der an der Börse gehandelt wird. Der jährliche
Umsatz beträgt etwa 20 Mrd. £.

Dabei haben sich in den 80er Jahren die Gewichte erheblich verschoben. Der
Marktanteil der Property Companies bei den Neuinvestitionen ist deutlich gestiegen.
Dies geschah u.a. unter dem Einfluß der Zunahme ausländischer Investoren. Am
bekanntesten ist das Beispiel Canary Wharf, wo allein rund 1 Mio. qm Büroflächen
errichtet werden. Diese Investitionen wurden fast völlig von den Property Companies
finanziert. Die institutionellen Anleger konzentrieren sich auf die etablierten und
weniger riskanten Standorte. Seit Beginn der 80er Jahre hat der Anteil der Immobilien-
gesellschaften, die sich überwiegend aus Bankdarlehen und der Ausgabe neuer Aktien
finanzieren, dramatisch erhöht ist, während die institutionellen Investoren ihre Immobilieninvestitionen seit Beginn der 80er zunächst real konstant hielten. Seit Mitte der 80er
Jahre sind die Investitionen sogar real gesunken. D.h. der Immobilienboom seit 1985
war im wesentlichen ein Boom, der durch die Property Companies hervorgerufen und
von Banken finanziert wurde.

Hoher Marktanteil der «Amateurinvestoren» in der Bundesrepublik
Aus den Unterschieden der Investoren in Großbritannien und der Bundesrepublik
ergeben sich erhebliche Verhaltensunterschiede am Markt. Ein britischer Experte, der
lange Jahre in Deutschland gearbeitet hat, formulierte die Unterschiede wie folgt:

Der Markt für Mietobjekte im Einzelhandels- und Bürosektor wird in der Bundesrepublik sehr stark durch den hohen Anteil privater Einzelvermieter bestimmt. Bei
diesen Einzelvermietern kann man, ohne daß dies in den Marktanteilen quantifizierbar
ist, verschiedene Typen unterscheiden.

- Auf der einen Seite gibt es die «Traditionseigentümer», die Objekte in der Familie zum Teil ererbt haben und sie seit langen Zeiten halten. Diese Eigentümer haben durch Traditionen beeinflußte Vorstellungen darüber, wie ihr Objekt genutzt werden sollte und welche Mieter zu ihnen passen (Prestige, Sektor, Art der Außendarstellung, etc.). Die Mietanpassungenwerden zum Teil bezogen auf die einzelnen Mieter vorgenommen. Es bestehen wenig Vergleichsmöglichkeiten zu anderen Objekten. Vor allem wird keine kalkulatorische Strategie der Verwertung vorgenommen. Die Verkaufsoption taucht nicht als systematische Verwertungsstrategie auf. In der Regel werden solche Objekte von Hausverwaltungen bewirtschaftet. Es bestehen oft über lange Jahre Geschäftsbeziehungen. Konkurrenten von außen können selbst mit dem Argument, daß sie sehr viel höhere Erträge erwirtschaften würden, in diese Vertragsbeziehungen nicht «einbrechen». Kommt es zu Veräußerungen, dann entstehen diese Veräußerungen stärker aus persönlichen Motiven als aus einer Portfolio-Optimierungs Strategie.
- Neben diesen Traditionseigentümern gibt es Personen mit sehr großen Vermögen, die selbst Immobilien erwerben. Zum Teil werden ererbte Vermögen ergänzt. Diese Personen sind wirtschaftlich sehr erfolgreich und von daher stärker an einer professionellen Bewirtschaftung interessiert. In unterschiedlichen Abstufungen entstehen hier selbständige Immobilienunternehmen dadurch, daß sehr große Vermögen oft von eigens dafür eingerichteten Verwaltungen bewirtschaftet werden. Hier gibt es Ähnlichkeiten zum amerikanischen Markt, wo ebenfalls sehr wohlhabende Investoren zum Teil eigene Immobilienverwaltungen «ausgründen».
  Gemessen am Marktanteil der privaten Eigentümer sind die Immobilienvermögen der Versicherungen Pensionskassen und andere Unternehmen relativ klein.[2]

Gründe für die Unterschiede

Die Erklärung für die unterschiedlichen Anbieterstrukturen sind nicht einfach zu finden. In der Diskussion mit verschiedenen Experten wurden folgende Faktoren genannt:
- In Großbritannien hat der Büromarkt in der Region London bzw. in Süd-Ost England eine überragende Stellung. An diesem Markt werden oft sehr große Objekte realisiert. Darüber hinaus sind großstädtische Märkte sehr komplex. Allein die Durchführung von Investitionen erfordert spezielles Wissen. Dies hat sein langem das Entstehen professioneller Investoren begünstigt.
- Private Investoren müssen in Großbritannien Steuer auf realisierte Wertsteigerung entrichten, d.h. der steuerliche Vorteil, den private Investoren in der Bundesrepublik gegenüber gewerblichen Investoren genießen, besteht nicht.
- Verluste aus Vermietung und Verpachtung privater Investoren konnten nicht mit anderen Einkunftsarten verrechnet werden.
- Die Versicherungen und Pension Funds haben sehr große Kapitalbeträge anzulegen. Anders als in der Bundesrepublik werden Pensionsverpflichtungen nicht direkt von Unternehmen eingegangen, die einen Pool zur Absicherung solcher Forderungen bilden. Die Pensionsrückstellungen werden generell auf Pensionskassen außerhalb der Unternehmen ausgelagert.[3]
- Der große Anteil der ausländischen Investoren in London hat das Wachstum der Property Companies begünstigt. In vielen Fällen gibt es Mischformen zwischen Developer und Property Company. Zum Teil werden Objekte für Großanleger erstellt und von diesen finanziert. Zum Teil werden Objekte zunächst in eigene Anlage-

vermögen genommen und anschließend veräußert.
- Die Unternehmen bilanzieren nicht zu Anschaffungswerten sondern zu Zeitwerten. Allein dadurch wären Immobilienaktiengesellschaften in der Bundesrepublik am Markt schwer bewertbar. Sie müßten eigene, freiwillige Zeitwertermittlungen vornehmen. Gegenwärtig gründet die DG-Bank in Frankfurt eine Grundstücksaktiengesellschaft. Der Erfolg dieses Versuchs wird von Experten bezweifelt. Neben der Risikoscheu deutscher Vermögensanleger und der Unsicherheit, welche Wertentwicklung Immobilienaktiengesellschaften nehmen werden, spielen hier auch Fragen der Bewertung der Aktien eine Rolle. Dies ist ein weiterer Faktor, der erklärt, daß in der Bundesrepublik Aktiengesellschaften am Immobilienmarkt kaum attraktiv sind.[4]

Folgerungen für das Investitionsverhalten und die Bewirtschaftungsform am Immobilienmarkt

Immobilienanlagen werden in Großbritannien sehr viel stärker als in der Bundesrepublik alsOptimierungsentscheidungen im Rahmen eines professionellen Portfoliomanagement angesehen. Britische Experten vertreten die These, daß auch die deutschen professionellen Anleger sich sehr viel traditioneller verhalten als englische Anleger und z.B. historisch gewachsene Immobilienportfolios über lange Zeit unverändert beibehalten. In Großbritannien werden dagegen die Renditen von Objekten ständig überprüft und mit anderen Bereichen verglichen. Falls sich bessere Alternativen zeigen, ist man bereit, daß jeweilige Objekt zu veräußern. Dies erklärt u.a. das extrem hohe Maß an Transaktionen. In den 80er Jahren ist die durchschnittliche Halteperiode bei institutionellen Investoren nach Auskunft von Experten auf rd. 4 Jahre abgesunken. Die Anleger kümmern sich relativ wenig um die Bewirtschaftung der Objekte. Mietverträge werden in der Regel auf 25 Jahre abgeschlossen, die Mieter übernehmen die vollen Bewirtschaftungskosten einschließlich der Steuerzahlungen für das Objekt. Die Mieten werden im 5jährigen Turnus an die Marktentwicklung angepaßt. Dabei wird vom Royal Institute der Chartered Surveyors jeweils ein Arbitrator ernannt, der die Vergleichsmiete festlegt. Die Unternehmen wissen regelmäßig, wie hoch die Äquivalenzmiete am Markt ist, wenn die letzte Mietanpassung einige Jahre zurückliegt, d.h. bei Veräußerungen werden diese Mieterhöhungsmöglichkeiten systematisch mit berücksichtigt.

Die Kontrolle bzw. die Verhandlungen mit den Mietern und die Festlegung der Details der Vermietung wird von Real Estate Agenturen oder von den Chartered Surveyors vorgenommen. Sie vertreten die Interessen der Eigentümer, die ihre Verwaltungs- und Bewirtschaftungsaufgaben weitgehend delegieren. In der Umkehrung folgt aus dieser Regelung, daß die Nettoerträge der Bewirtschaftung sehr viel eindeutiger ermittelt werden können, als in der Bundesrepublik, was wiederum die Fungibilität von Immobilienanlagen erhöht. Jeder Investor kann innerhalb kurzer Zeit die Ertragssituation eines Objektes beurteilen.

Aus den Verhaltensweisen erklären sich die Marktverläufe. Wie die Grafik 9 auf S. 16 zeigt, hat der Anteil der Immobilien an den Vermögensanlagen der institutionellen Anleger zu Beginn der 80er Jahre ein Maximum erreicht. Dies war weniger aus den damaligen Renditesituationen sondern aus längerfristigen Strategien, die u.a. mit den Inflationserwartungen zusammenhängen zu erklären. Im Verlauf der 80er Jahre haben die Immobilienanlagen ständig abgenommen. Dies erklärt sich aus dem relativen Abfall der Renditen. Gestiegen sind im Verlauf der 80er Jahre vor allem die Auslandsinvestitionen und die Investitionen in Beteiligungswerten. Statistisch noch nicht erfaßt sind die jüngsten Trends. Wie Experten versichern, haben die institutionellen Anleger seit 1989 im größeren Stil Objekte veräußert, insbesondere an ausländische Investoren. Dies hat zur Folge, daß der jetzige Renditeverfall und der Rückgang der Rentabilität vor allem von ausländischen Investoren und Banken getragen wird.

Unterschiede der Marktorganisation

Geringe Transaktionskosten

Die Transaktionskosten werden bei Büroobjekten mit etwa 2% angegeben. Hier spielt eine Rolle, daß es keine Grunderwerbsteuer gibt. Aber auch die Gebühren der Makler sind niedriger als bei uns. Anstelle der Notarkosten fallen die Rechtsanwaltskosten an. Nach Auskunft von Experten, die in beiden Märkten tätig sind, entsprechen sich Notarkosten und englische Rechtsanwaltskosten weitgehend. Neben den relativ geringen Kosten spielen die routinemäßig leicht verfügbaren Informationen eine erhebliche Rolle. So erhalten die potentiellen Käufer regelmäßig standardisierte und deshalb leicht vergleichbare Gutachten über die Ertragssituation der von ihnen erworbenen Gebäude. Sie können leicht Vergleiche zu anderen Objekten ziehen. Das heißt neben den geldlichen Organisationskosten wird auch der Zeit-, Entscheidungs- und Analyseaufwand möglichst niedrig gehalten.

Die Rolle der Surveyor

Die Surveyor als eine eigene Berufsgruppe haben eine sehr bedeutsame Funktion am englischen Immobilienmarkt. In ihrer Ausbildng müssen sie, nach einem akademischen Grad, eine dreijährige zusätzliche Fachausbildung vornehmen, die aus einer Kombination von Praxis und Hochschulausbildung besteht. Dabei gibt es unterschiedliche Spezialisierungsrichtungen. Es besteht die Möglichkeit, sich stärker juristisch-wirtschaftlich zu entwickeln, oder stärker technisch-bauökonomisch (Quantity Surveyor, Quality Surveyor). Dem entsprechend treten die Surveyor in ganz unterschiedlichen Zusammenhängen auf:
- Bei allen Immobilientransaktionen sind Bewertungen vorzunehmen. Diese werden ausschließlich von unabhängigen Surveyorn durchgeführt.
- Alle großen Surveyorfirmen bewirtschaften für ihre Kunden Objekte. Dabei werden ihnen die Kontroll- und Interessenvertretungsfunktionen der Eigentümer fast vollständig übertragen.
- Bei Grundstücksumsätzen treten die einzelnen Surveyor als Interessenvertreter einer Partei auf, d.h. der Verkäufer beauftragt einen Surveyor mit der Wahrnehmung seiner Interessen. Der Käufer wird von einem anderen Surveyor vertreten. Die Interessenvertretung beinhaltet deshalb eine genaue Prüfung der Objekte und eine Einschätzung der künftigen Rentabilität.
- Bei der Durchführung von Bauprojekten treten die Surveyor vielfach als Projektentwickler auf, d.h. sie klären die Grundstücksfragen, beantragen Baugenehmigungen und verhandeln mit den Kommunen. Dabei kommt ihnen ihre regionale Verwurzelung zugute. Vor allem Investoren, die von außen oder aus dem Ausland kommen, stützen

sich auf solche lokalen Dienstleistungen. In der Projektdurchführung gibt es spezialisierte Surveyor bzw. Abteilungen innerhalb großer Unternehmen, die eine Kostenkontrolle vornehmen oder auch vorher die Kostenschätzungen der Bauunternehmen überprüfen.

- Surveyor stellen Marktinformationen zur Verfügung. Sie beobachten den Grundstücksmarkt ständig und veröffentlichen eigene Rendite- und Umsatzstatistiken. Allerdings verweisen Anleger darauf, daß sie es vorziehen unabhängige Consultants für Markt- und Standortbewertungen heranzuziehen, weil die reinen Beratungsunternehmen keine Interessen hinsichtlich Realisierung oder Nicht-Realisierung von Projekten haben.

Von Experten, die sowohl in Großbritannien wie in der Bundesrepublik tätig sind, wird auf einen Punkt mit hohen Nachdruck verwiesen: Die Tatsache, daß Surveyor nicht lediglich Makler sind, die ihre Erlöse aus Provisionen bei Umsätzen erzielen, sondern eher fachlich spezialisierte Dienstleistungsanbieter (Beratungen, Bewirtschaftungen, Projektentwicklungen), trägt erheblich zur Qualität der Leistungen und Qualifikation des Berufsstandes bei. Auf der Einnahmeseite der Surveyor und Real Estate Agenturen spielen die Dienstleistungseinnahmen eine sehr große Rolle. Surveyor haben zu großen Investoren oft lange vertragliche Bindungen und Beziehungen, d.h. sie beraten bei Investitionsplanungen, vermarkten fertiggestellte Objekte und bewirtschaften sie. Aus solcher vertrauensvollen, langfristigen Zusammenarbeit entstehen andere Interessenorientierungen. Es bildet sich vor allem ein Leistungswettbewerb als Qualitätswettbewerb von Dienstleistungen.

In der Bundesrepublik deuten die Trends der letzten Jahre darauf hin, daß sich eine solche Rolle der großen Makler auch allmählich herausbildet. Typisch ist, daß mehrere Makler eigene Hausverwaltungsorganisationen ausgegründet haben. Einzelne Makler treten auch schon als Projektentwickler auf.

Standardisierte Mietverträge

Als ein weiterer Vorteil des englischen Immobilienmarkts werden die standardisierten Mietverträge angesehen. In der Regel werden Verträge auf 25 Jahre geschlossen. Es kommt in regelmäßigen Abständen von 5 Jahren zu einer Mietanpassung. Anpassungen werden vertraglich in der Regel nur nach oben vorgenommen. Die Mietfestsetzung wird durch Surveyor vorgenommen. Dadurch ergeben sich relativ wenig Konflikte und Streitigkeiten bei der Mietanpassung. Voraussetzung dafür ist allerdings auch, daß eine hohe Markttransparenz besteht. Die standardisierten Verträge erlauben es darüberhinaus die Rentabilität von Objekten sehr rasch zu beurteilen. Für di Investoren ist von Bedeutung, daß Mieter, die einen Vertrag vorzeitig kündigen, oft nicht aus der vertraglichen Verpflichtung zur Zahlung der Miete entlassen werden. Falls der Nachmieter eine geringere Bonität hat und seinen vertraglichen Verpflichtungen nicht voll nachkommt, haftet der Vormieter für die Restlaufzeit seines vorzeitig gekündigten Vertrages.

Marktinformationssysteme

Angesichts der hohen Transaktionen und des hohen Bedarfs an Informationen, sowie der hohen fachlichen Spezialisierung, verwundert es nicht, daß am Markt leistungsfähige Informationssysteme angeboten werden, die Umsätze, Renditen, Mieten und Gesamtrentabilität darstellen. Es gibt mehrere Gesellschaften, die von den Großeigentümern

Verkehrswerte und Bewirtschaftungsdaten der Einzelobjekte erhalten und aus diesen Einzelinformationen Renditeindikatoren für Regionen und Sektoren fortlaufend veröffentlichen. Diese Informationen können käuflich erworben werden. Die Tatsache, daß sich solche Informationssysteme kommerziell selbst tragen deutet auf eine hohe Nachfrage nach solchen Informationen hin. Die Miet- und Wertsteigerungsrenditen werden jeweils zu einem Gesamtindikator zusammengefaßt.

Aus der ausgeprägten Arbeitsteilung zwischen Vermögensanlegern und den Managementorganisationen folgt, daß in Großbritannien die Dienstleistungen, die sich um die Immobilienanlage ranken, sehr viel besser entwickelt sind als in der Bundesrepublik. Allein die hohen Transaktionen lassen sich mit Erfolg nur abwickeln, wenn eine Markttransparenz entsteht und entsprechende Dienstleistungskapazitäten jeweils dafür sorgen, daß diese Transaktionen mit geringen Kosten möglich werden. Wie dargestellt betragen die Transaktionskosten nach Auskunft von Experten etwa 2% des Kaufpreises. Aus dem großen Umfang der Transaktionen folgt, daß für hochspezialisierte Makler Verwaltungs- und Beratungsdienstleistungen ein größerer Markt besteht. Objekte müssen z.B. jeweils neu bewertet und begutachtet werden. Dafür wurden Standardverfahren und Standardmaßstäbe entwickelt. Bedeutsam ist, daß die Marktinformationen systematisch gesammelt und in kondensierter Form veröffentlicht wurden. Das hohe Volumen führt zu reichhaltigen Daten. Aufgrund der starken Nachfrage können diese Daten wiederum mit einem hohen Aufwand analysiert und aufbereitet werden.

Markttransparenz über Bautätigkeit

Die amtlichen Daten erlauben detailliertere Informationen. Anders als in der Bundesrepublik wird ein Bauantrag nicht als privates, geschütztes Datum angesehen. Bauanträge gelten als öffentlich zugängliche Informationen. Das Interesse der Allgemeinheit, frühzeitig Informationen über Investitionsabsichten und Pläne zu erhalten, wird höher eingeschätzt als das Interesse des Investors diese Pläne nicht bekannt werden zu lassen. Dabei wird auch darauf verwiesen, daß zwar das Interesse des einzelnen Investors an der möglichst langen Geheimhaltung seines individuellen Projektes besteht, daß der gleiche Investor jedoch zur rationalen Begründung und Kalkulation seines Projektes ein Interesse an Information über benachbarte Konkurrenzprojekte hat. Angesichts dieser Situation müssen rationale Investoren akzeptieren, daß ihr Informationsbedürfnis nur befriedigt werden kann, wenn die eigene Investition ebenfalls in einem gemeinsamen Datenpool ausgewertet, analysiert und bekannt gemacht wird. Diese Einstellung hat zur Folge, daß Statistiken über Bauanträge veröffentlich werden. Daneben werden aber auch in Gebieten mit hoher Investitionsdichte bzw. in Gebieten mit hohem Investitionsinteresse zum Teil von privaten Firmen parzellenscharfe Bauantrags- und Genehmigungsdaten veröffentlicht.

Die Basisdaten für solche Investitionskarten stammen aus der Bautätigkeitsstatistik, die auch in ihren Einzeldaten öffentlich zugänglich ist. Von allen Beteiligten wird die erhöhte Markttransparanz als unbedingt erforderlich angesehen, um Investitionsentscheidungen möglichst rational zu treffen. Es bleiben trotz hoher Markttransparenz und guter Informationen noch genügend Kalkulations- und Planungsrisiken, wie die starken zyklische Schwankungen im Immobilienmarkt zeigen. Dabei wird von Investoren immer wieder darauf verwiesen, daß diese zyklischen Schwankungen fast regelmäßig durch exogene politische Einflüsse zustande kamen und nicht als endogene Marktphänomene angesehen werden können.

Unterschiede bei der Abwicklung von Transaktionen

In Großbritannien gibt es kein Grundbuch, das mit den staatlichen Grundbuchleistungen in der Bundesrepublik vergleichbar wäre. Dies hat zur Folge, daß die entsprechenden Informationen privat beschafft und bereitgestellt werden müssen. Bei Grundstücksverkäufen setzen die Parteien regelmäßig Rechtsanwälte ein, die wiederum durch «title search» die Eigentumsrechte feststellen. In diesem Punkt sind die Rahmenbedingungen in der Bundesrepublik günstiger, weil Eigentums- und Beleihungsverhältnisse sowie sonstige Belastungen der Grundstücke ohne Zeitverzögerung aus dem Grundbuch entnommen werden können. Dementsprechend können Immobilientransaktionen in der Bundesrepublik notfalls innerhalb von Stunden abgewickelt werden, wenn die Finanzierungsfragen und sonstige Vertragskonstruktionen geklärt sind. In Großbritannien werden zunächst Verträge ausgetauscht (exchange of contracts). Erst nach der Prüfung der Verträge kommt es zu einem abschließenden Vertragsabschluß (completion). Die Experten verweisen jedoch darauf, daß in der Bundesrepublik für Käufer und Verkäufer jeweils das Risiko besteht, daß bis zur Unterzeichnung keine verbindlichen Verpflichtungen bestehen. Anbieter und Nachfrager können bis zur Unterzeichnung beim Notar ihre Positionen ändern. Nicht selten werden in letzter Minute neue Nachfrager, die höhere Preise bieten, präsentiert. Dies wird von britischen Experten, die in der Bundesrepublik tätig sind, als unerträglich angesehen. Nach ihrer Auffassung gehört es zum ordentlichen Geschäftsverkehr, daß Angebote und die entsprechenden Entscheidungen der Nachfrager verbindlich sind. An dieser Kritik wird deutlich, daß professionelle Immobilieninvestoren den besonderen Schutz, den Anbieter und Nachfrager in der Bundesrepublik gegenüber übereilten Entscheidungen genießen, nicht für erforderlich halten. Angesichts des festgefügten und eingespielten Systems erscheint es jedoch kaum möglich und sinnvoll hier gravierende Änderungen vorzunehmen. Es könnte allerdings sinnvoll sein, im Verkehr zwischen Geschäftsleuten ein höheres Maß an Verbindlichkeit bei Angeboten und Nachfragegeboten gelten zu lassen.

Zur Erklärung der starken Zyklen

Im Vergleich zur Bundesrepublik entstehen in Großbritannien an den Immobilienmärkten seit langem extreme zyklische Schwankungen. Die Schwankungen rufen erhebliche Störungen der Wirtschaftsentwicklung hervor, sie schlagen sehr stark auf die Bautätigkeit zurück. Sowohl zu Beginn der 70er Jahre, wie Ende der 80er Jahre kam es zu dramatischer Überproduktion mit einem anschließendem Zusammenbruch der Märkte. Angesichts der hohen Markttransparenz und der ständigen Marktanalysen, die vorgenommen werden, stellt sich die Frage, wie solche Zyklen möglich sind, warum aus der Vergangenheit nicht gelernt wird. Zu diesem Thema gibt es eine ausgedehnte Fachdiskussion.[5] Dabei stößt man auf die üblichen Erklärungen:
- Lange Planungs- und Vorbereitungsfristen.
- Die Projekte werden in Situationen in Angriff genommen, in denen Renditen steigen und Erwartungen günstig sind. Die Realisierungsphasen fallen in Perioden, die von ganz anderen ökonomischen Randbedingungen geprägt sind.
- Die Realisierungsfristen können von den Investoren nicht hinreichend kontrolliert werden. Als Ergebnis entsteht der Zwang auch Projekte zu realisieren, bei denen lange vorher absehbar ist, daß die Ertragsaussichten sich dramatisch verschlechtert haben.[6]

In London lieferte die Konkurrenz zwischen den Docklands und der City einen besonderen Grund für das Auftreten zyklischer Erscheinungen. Dabei wurde die Investitionstätigkeit in den Docklands durch spezielle Steuervorteile angeheizt. (Enterprise zone: Canary Wharf – Sofortabschreibung, 10 jährige Grundsteuerbefreiung) Nachdem n den Docklands mehrere Mio. qm Bürofläche geplant oder in Angriff genommen wurden, entstand in der City die Befürchtung eines Bedeutungsverlusts. Darauf wurde mit einer generellen Erhöhung der Baudichte und damit einer planerischen Expansion der Bautätigkeit reagiert.[7]

Diese Situation, die sich z.b. auch in der Konkurrenz zwischen den Großprojekten Broadgate und Canary Wharf widerspiegelt, hat das overshooting der Investitionstätigkeit erheblich begünstigt. Daraus wird deutlich, daß der gegenwärtige Boom durch eine besondere Kombination von Faktoren angeregt wurde. Die Fertigstellungsphase fällt zusammen mit drastischen Maßnahmen zur Inflations-bekämpfung und einer Stabilisierungskrise, die zwar nicht das Ausmaß von 1982 erreicht, die jedoch auf mehrere Jahre die Wachstumsraten senkt. Hintergrund dieser Stabilisierungspolitik war u.a. ein drastischer Abfall der Sparquoten und in der Umkehrung ein ausgesprochener Konsumboom der die Inflationstendenzen begünstigte.

Für die Bundesrepublik bleibt die Folgerung, daß die Internationalisierung der Märkte die Tendenzen zur Instabilität an den Immobilienmärkten verstärken werden. Eine höhere Markttransparenz kann diesen Tendenzen z.T. begegnen. Extreme Schwankungen, wie sie in Großbritannien aufgetreten sind, gehen jedoch auch zurück auf allgemeine starke Schwankungen der Wirtschaftsentwicklung (Schwankungen der Sparquote, die weit höher sind als in der Bundesrepublik, starke Schwankungen der Inflationsraten und Wachstumsraten und damit auch der Erwartungen).

Auch in der Bundesrepublik kann man ein Zusammenfallen mehrerer Sonderfaktoren nicht ausschließen. So ist durchaus wahrscheinlich, daß unter den gegenwärtigen Sonderbedingungen Berlins innerhalb von 5 bis 6 Jahren eine extreme Überinvestition am Büromarkt entsteht. Die Mischung der Faktoren ist hierfür ausgesprochen typisch. Dabei muß man davon ausgehen, daß in Berlin ein großer Teil der Büroinvestitionen von internationalen Anlegern und Firmen finanziert und durchgeführt werden. Es bestehen keinerlei Erfahrungen mit den langfristigen Wirtschaftstrends der Stadt. Die hohen Immobilienpreissteigerungen rufen die Tendenz hervor, sich zunächst selbst zu verstärken, weil durch Kauf und Verkauf hohe Gewinne erzielbar sind. Ein solcher sich selbst ernährender Immobilienboom birgt per se den Keim zum Zusammenbruch in sich. Dieser Zusammenbruch wird dann kommen, wenn hohe Fertigstellungen in der Phase nach 1995 nicht auf eine entsprechend gestiegene Nachfrage nach Büroflächen stoßen.

Fazit

Die Frage optimaler Rahmenbedingungen wird gegenwärtig in der Bundesrepublik kaum diskutiert. Steuerrecht- und Bilanzfragen sind eingeübt und etabliert. Die Eigentümer und Mieter haben kaum ein Interesse an verschärftem Wettbewerb. Von daher ist die Situation in jeder Hinsicht stabil. Es könnte allerdings sein, daß durch verschärften internationalen Wettbewerb und durch internationale Kontakte Marktinstitutionen und Marktabläufe verändert werden müssen oder sich verändern. Gegenwertig steigen die grenzüberschreitenden Immobilieninvestitionen erheblich an. Auch Dienstleistungsunternehmen und Finanzierungsinstitutionen betätigen sich immer stärker auf internationalen Märkten. In der Bundesrepublik wurde dies besonders sichtbar daran, daß Großinvestitionen z.b. der Messeturm in Frankfurt von ausländischen Investoren errichtet wurden. Am Markt für Dienstleistungen sind inzwischen fast alle großen englischen Maklerfirmen in verschiedenen Großstädten präsent. Diese gengenseitige Marktdurchdringung und die erhöhten Marktkontakte werden Anpassungen an veränderte Rahmenbedingungen und Standortpotentiale erleichtern helfen.

4

Die Kommunen und die Entwicklung des Immobilienmarktes

Partnerschaftliche Grundstücksentwicklung

Das englische Grundstücksrecht gewährt Investoren eine hohe Flexibilität bei der Realisierung, Finanzierung und Bewirtschaftung von Immobilieninvestitionen. Wichtig ist dabei, die sehr kreative Handhabung von langfristigen Leaseverträgen. Je nach Ausgestaltung ähneln Leaseverträge mehr einem schuldrechtlchen Mietvertrag oder einem dinglichen Erbbaurecht. In der Regel werden 25 jährige Mietverträge vereinbart, bei denen die Mieter (Inhaber von Leaserechten) für Steuern und Abgaben aber auch für die Instandhaltung der Gebäude zuständig sind. Leaserechte für Grundstücke werden aber auch sehr häufig über 125 Jahre oder noch länger gewährt. Sie sind verkaufbar und vererbbar und nähern sie sich damit einem Erbbaurecht.

Die Kommunen, aber auch private Investoren und Entwicklungsgesellschaften nutzen Leaserechte in unterschiedlicher Weise, um Erträge, Rechte und Pflichten bezogen auf ein Grundstück aufzuspalten. Im Einzelfall hängt die gefundene Lösung von den Zielen ab, die mit der Grundstücksbewirtschaftung erreicht werden soll. So können die Gemeinden das Entwicklungsrisiko für Investoren senken, indem sie Erbbaurechte zu anfänglich niedrigen Verzinsungen vergeben. Wollen Sie dagegen auf die Nutzung und Bewirtschaftung des Objekts Einfluß nehmen, können sie das Grundstück verkaufen und sich am bebauten Objekt ein Leaserecht vertraglich vorbehalten. Dies würde dann z.B. eine Einflußnahme auf die Nutzungsmischung ermöglichen. Sollen möglichst hohe fiskalische Erträge erzielt werden, unabhängig von der Art der gefundenen Nutzung, dann können joint venture Lösungen sinnvoll sein, bei denen das Grundstück in eine gemeinsame Entwicklung eingebracht wird und die Gemeinde auf Dauer mit einem bestimmten Prozentsatz an den Erträgen beteiligt wird. Die Art der Lösungen ist sehr vielfältig. Hier seien nur einige typische Beispiele aufgelistet:
- Der Freeholder (Volleigentümer) möchte seine Nutzungsrechte und Dispositions-möglichkeiten in der Nutzung aufrecht erhalten. D.h., er möchte z.B. sein Einzelhandelsgeschäft in dem eigengenutzten Gebäude weiterbetreiben. Gleichzeitig will er jedoch über die stillen Reserven des Grundstückseigentums verfügen, um daraus weitere Expansionen zu finanzieren. Um dieses Ziel zu erreichen, kann das Freeholdrecht verkauft werden, wobei der Käufer dem Verkäufer sofort ein Leaserecht/ Erbbaurecht zurückgewährt (Leaseback). Dabei können die Erbbauzinsen z.B. ex ante festgelegt werden, um den Käufer eine feste Rendite zu gewähren. Der Verkäufer und Leasevertragnehmer hat weiterhin die volle Chance der Grundstücks-verwertung und kann die vollen Erträge erzielen. Er kann sich auch eine Option zur

222

Fortsetzung des Leasevertrages aushandeln. Der Langfristinvestor erzielt eine vereinbarte Rendite und verfügt darüber hinaus über die Wertsteigerungserwartungen und die künftigen Gewinnerwartungen.

- Ein Grundstückseigentümer möchte die Entwicklung seines Grundstückes nicht selbst vornehmen. Er schließt einen Vertrag mit einer Entwicklungsgesellschaft. Er will Grundstückseigentümer bleiben und vereinbart eine Aufteilung der Erträge entsprechend eines Anteilsverhältnis des Grundstückswertes zum Investitionswert. Der Investor ist nicht an einer langfristigen Bewirtschaftungen interessiert. D.h., der Grundstückseigentümer schließt gleichzeitig einen Leasevertrag über das Gebäude ab und übernimmt die Bewirtschaftung. In diesem Fall sind wiederum verschiedene Lösungen möglich. So kann der Investor eine garantierte Verzinsung seiner Investition erhalten und damit keinerlei Risiko übernehmen. Der Grundeigentümer und Mieter des Gebäudes übernimmt das volle Risiko. Es sind aber auch Regelungen denkbar, nach denen der Investor eine Mindesrendite garantiert erhält und die über-schießenden Sonderzusatzgewinne nach bestimmten Schlüsseln verteilt werden.

- Die Kommunen nutzen solche Regelungen häufig, um risikoreiche Investitionen anzuregen. So können die Investoren Grundstücke zunächst preisfrei erhalten. Die Kommunen verzichten auf Erträge und verringern damit den Kapital- und Finanzier-ungsbedarf für risikoträchtige Investitionen. Stellt sich nach einiger Zeit heraus, daß die Investitionen ein Erfolg waren, dann werden aufgrund vertraglicher Verein-barungen die Kommunen an diesen Erträgen beteiligt. Dabei werden in der Regel progressive Beteiligungen estgelegt. Für den Investor führt dies dazu, daß sich die Bandbreite der eigenen Nettoerträge einengt. Bei ungünstigem Investitionsergebnis sinken die Erträge nicht so stark ab, weil dann zunächst die Kommunen auf Gewinne verzichten. Wird die Investition dagegen sehr gewinnträchtig, dann werden die Kommunen progressiv beteiligt mit dem Ergebnis, daß der Nettoerfolg für den Investor schwächer ausfällt. Eine solche Strategie ist aus der Sicht des Investors durchaus rational, weil sie Risiken vermindert und vor allem Absicherung gegenüber dem Negativfall eines Kapitalverlusts oder verlustreicher Investitionen bietet.

Die Beispiele zeigen, welche Bandbreite partnerschaftlicher Immobilieninvestitionen, gestützt vor allem auf die Überlagerung mehrerer Rechte in der Praxis anzutreffen sind. Dabei werden solche Regelungen nicht nur von den Kommunen sondern auch von privaten Investoren immer häufiger eingesetzt. Eine Rolle spielt dabei, daß in den letzten Jahren vor allem Großunternehmen ihre Grundstückswirtschaft erheblich rationalisiert haben. Bloße Verkaufsstrategien werden damit seltener. Die Groß-eigentümer von Immobilien werden immer häufiger zu Grundstücksentwicklern bzw. zu Langfristinvestoren am Immobilienmarkt, da bei steigender Grundstücksknappheit hier auf Dauer erhebliche Erträge erzielbar sind.

Angesichts der hohen Risiken aufgrund spezieller örtlicher Lagefaktoren sind Eigentümer oft interessiert, die Risiken vor allem von Pionierinvestoren in neuen Gewerbegebieten oder in Erneuerungsgebieten zu senken. Dabei werden entwick-lungsorientierte Finanzierungs- und Vertragskonstruktionen gewählt. In der Regel läuft dies darauf hinaus, daß der Grundstückseigentümer auf Sofortertäge verzichtet und an langfristigen Entwicklungserfolgen beteiligt wird. Dafür eignen sich partnerschaftliche Investitionen, bei denen Eigentum und Leaserechte in Kombination benutzt werden.

In der Bundesrepublik spielen solche Techniken quantitativ eine geringere Rolle. Kommunen und andere öffentliche Eigentümer nutzen Erbbaurechte oder Joint Venture

Regelungen seltener um zur Entwicklung von Grundstücken oder Stadtteilen beizutragen. In der jüngsten Zeit mehren sich jedoch die Anzeichen, daß auch solche Techniken häufiger auftreten.

Unabhängig von den entwicklungsorientierten oder ertragsorientierten Motiven für solche Rechtskonstruktionen scheint es ohnehin sinnvoll, daß die Kommunen und Großeigentümer von Grundstücken ihre langfristigen Strategien überdenken. Angesichts der wachsenden Flächenansprüche, der wieder steigenden Bevölkerung und der gesunkenen Bereitschaft der Kommunen, auf die steigende Nachfrage mit einer elastischen Baurechtspolitik zu reagieren ist abzusehen, daß Grundstücke vor allem in günstigen Lagen immer knapper werden. Das tatsächliche Verhalten läuft auf eine Rationierung von Baurechten hinaus. Das bedeutet, daß vor allem bestehende Baurechte Knappheitsrenten erhalten, die nicht durch Wettbewerb beseitigt werden können. Verkaufsstrategien erscheinen vor diesem Hintergrund aus mehreren Gründen suboptimal:

– Grundstückseigentümer verlieren auf Dauer die Einflußmöglichkeit auf die Nutzung und Umnutzung von Grundstücken. Angesichts der wachsenden Nutzungskonflikte kann sich diese Einflußminderung, gerade in Großstädten, als schädlich für die Stadtentwicklung erweisen.

– Unter fiskalischen Gesichtspunkten werden Verkaufsstrategien ebenfalls suboptimal, weil Beteiligungen an langfristigen Erträgen wahrscheinlich deutlich günstigere Ergebnisse erbringen.

– Schließlich werden vor allem in den Kernstädten großer Verdichtungsräume die Investitionsprojekte immer komplexer. Es kommt immer häufiger zu Verzahnungen von öffentlichen und privaten Investitionen. Hier eignen sich gemeinsame Grundstücksverwertungen durch Aufspaltung von Grundstücksrechten, um wirtschaftlich und juristisch günstige Lösungen zu finden. In Hafengebieten sind solche Kombinationen schon immer üblich gewesen. Die Gründe, die dort zu entsprechenden Rechtsaufteilungen geführt haben, treffen immer mehr auf entrale Grundstücke zu.

Kommunale Strategien gegenüber Wertsteigerungen aufgrund von Planungsentscheidungen

Hintergrund

Großbritannien ist das Heimatland des Planungswertausgleichs. Noch während des Krieges hat die Uthwatt-Kommission ein Konzept vorgelegt, nach dem alle künftigen Planungswertsteigerungen an Grundstücken sozialisiert werden sollten. Die Kommission formulierte die Position, daß Bebauungsrechte bzw. Rechte zur intensiveren Ausnutzung von Grundstücken nicht als allgemein inhärenter Bestandteil des Eigentumsrechts gelten können. Dementsprechend sollten alle Nutzungsintensivierungen als von der Gemeinschaft gewährte Rechte gelten, deren Wertsteigerung der Gemeinschaft zufallen sollte.

Jede Planung erzeugt Ungleichheit. Will man die Ungleicheit kompensieren, muß man eine rigorose Abschöpfung aller planungsbedingten Wertsteigerungen befürworten. Die seit dem 2. Weltkrieg in Großbritannien gestarteten Versuche zur Lösung des Problems sind weitgehend gescheitert. Zunächst wurde 1947 ein allgemeiner Planungswertausgleich eingeführt, der 1952 wieder abgeschafft wurde. 1967 kam es zu einer erneuten Einführung mit einem reduzierten Abschöpfungssatz. Auch diese Lösung hatte keinen Bestand. Später wurde eine Steuer auf realisierte Wertsteigerungen angewandt. Diese verschiedenen Ansätze machen deutlich, daß die Diskussion niemals zur Ruhe kam. Gegenwärtig herrscht der unsystematische Zustand, daß gewisse Wertsteigerungen bei ihrer Realisierung besteuert werden. Daneben können Kommunen und andere öffentliche Institutionen Grundstücke zum Teil zum Current Use Value, d.h. zum Wert in der historischen Nutzung enteignen und die Wertdifferenz bei Veräußerungen abschöpfen. Im Ergebnis führt die Enteignung damit zu einer Ungleichbehandlung.

Planning Gain und Developer Agreements

Unabhängig von theoretisch-systematischen Lösungen besteht nach dem Town and Country Planning Act 1990, Section 106 die Möglichkeit, Regelungen zu finden, nach denen die Eigentümer, Investoren und Begünstigten von Planungsentscheidungen sich verpflichten, an anderer Stelle Leistungen zugunsten der Kommune zu erbringen.

Es kommt jeweils zu Planning Agreements, bei denen der Begünstigte einer Planungsentscheidung sich verpflichtet, über die öffentlich-rechtlichen Planungsauflagen hinaus, im Rahmen der Entwicklung bestimmte Leistungen zugunsten der Kommune zu erbringen. In der Regel werden Leistungen übernommen, die sonst in die Aufgabenbereiche der Kommune gefallen wären. Planungsbedingte Wertsteigerungen werden auf diese Weise in gewissem Umfang zur Finanzierung öffentlicher Ausgaben beitragen.

Planning Gain-Vereinbarungen treten in Großbritannien typischerweise dann auf, wenn sehr komplexe Großprojekte realisiert werden, bei denen öffentliche und private Investitionen eng miteinander verzahnt sind, bei denen die Kommune Mitentwickler wird. Dadurch verlieren im Vergleich zur Planungswertsteigerungsabschöpfung die Vereinbarungen den Charakter einer Quasisteuer. Vielmehr werden sie Teil eines komplexen Vertragsbündels, in dem die privaten Investoren ganz unterschiedliche Leistungen übernehmen. Dabei spielt auch eine Rolle, daß die öffentlich-private Arbeitsteilung nicht so starr festgelegt ist wie in der Bundesrepublik. Von Fall zu Fall kommt es zu Grenzverschiebungen der Verantwortung (vgl. als Beispiel das King's Cross Projekt). Von den Investoren wird darauf verwiesen, daß die Kommunen die Grenzen der Arbeitsteilung zwischen Investor und Kommune immer wieder nach ihren Zwecken verschieben und den Investoren von Fall zu Fall größere Leistungen aufbürden (Straßenreparaturen im Zusammenhang mit Großprojekten, Modernisierung der Kanalisation, Neuanlage von Parkplätzen, usw.). Da es eine präzise Definition der möglichen Inhalte von Development Agreements nicht gibt, hängt das Ausmaß der Auflagen oder der Entwicklungen, die den privaten Investoren übertragen werden, von der jeweiligen Planungsmacht der Kommune ab. Handelt es sich um Grundstücke mit speziellen Lagevorteilen, die nicht durch andere Investitionen «wegkonkurriert» werden können, dann kann die Kommune durch hartnäckiges Verhandeln sehr hohe Nebenleistungen für sich erwirtschaften. Eine solche Position fehlt dann, wenn Konkurrenzinvestitionen günstiger realisiert werden können und dem anderen Projekt nicht gleichermaßen hohe Zusatzauflagen gemacht werden.

Development Impact Fees und ähnliche Konzepte in den USA
Die kommunalen Kompetenzen und auch die kommunalen Finanzen sind in den USA ganz anders organisiert als in Großbritannien. Immer wieder läßt sich jedoch beobachten, daß Analogien in der Planungsdiskussion auftreten. Dies hängt auch damit zusammen, daß gemeinsame Rechtstraditionen bestehen. Ähnlich wie in Großbritannien gibt es seit einiger Zeit in den USA eine ausgedehnte Fachdiskussion und zahlreiche Versuche die planungsbedingten Wertsteigerungen zugunsten der öffentlichen Hand abzuschöpfen. Hintergrund ist auch hier, daß Bebauungspläne den Charakter eines Wertpapiers zum Preis von Null haben. Dies hat auch in den USA immer wieder dazu geführt, daß Kommunen versuchen, zumindest einen Teil der im Zuge der Planung entstehenden Werte zur Finanzierung öffentlicher Leistungen heranzuziehen. In den USA ist eine solche Praxis noch sehr viel weitreichender. Es werden verschiedene Strategien verfolgt:

Bei den Linkage-Strategien geht es nicht nur um die Gewährung von Baurechten sondern ganz generell um Vorteile, die private Investoren aus Kooperationen mit der öffentlichen Hand haben. Die Investoren werden dann verpflichtet, als Gegenleistung gemeinnützige Investitionen oder Aktivitäten zu finanzieren. Beispiel: Gewährung von großzügigen Baurechten auf einem Bürogrundstück (zusätzliche Zahl von Stockwerken), gleichzeitig wird der Investor verpflichtet, in einer weniger attraktiven Gegend mit einem hohen Anteil von Minderheiten ein zusätzliches Bürogebäude zu errichten. Solche komplementären Verpflichtungen können sich auch auf den Bau von Wohnungen zugunsten bestimmter Bevölkerungsgruppen beziehen. Daneben werden aber auch unmittelbar mit der Investition zusammenhängende Infrastrukturmaßnahmen von dem privaten Investor übernommen (Parks, Straßenbau und Straßenreparaturen).

Leider sind diese Ad hoc-Gegenleistung in ihrem Wert schwer abzuschätzen. Beispiel: Werden Bürogebäude in einem risikoreichen Gebiet mit einem hohen Anteil von Minderheiten errichtet, kann dies u.U. überhaup kein «Opfer» für den Investor bedeuten, wenn die Investition auch in dem problematischen Stadtteil zu einem Erfolg wird.

Das zentrale Argument gegen die englischen und amerikanischen Methoden einer Abschöpfung von Planungsmehrwerten liegt auf der Hand: Es werden Investitionen angeregt, deren Ertragsminderung, die aus der ungünstigen Standortsituation oder aus anderen Randbedingungen entsteht, kaum abzuschätzen ist.

Als Folge dieser unsystematischen und von Fall zu Fall schwankenden Gegenleistungen wird in den USA seit längerem das Konzept der «Development Impact Fees» diskutiert. In sachlicher Nähe zu den Impact Fees stehen «Assessment Districts». Dies sind klar abgegrenzte Gebiete, die durch öffentliche Maßnahmen besonders begünstigt werden und die durch Sonderabgaben die begünstigen, die Leistungen in diesem Gebiet finanzieren (Investitionen tätigen).

Die Development Impact Fees sind zusätzlich zu den Investitionen zu finanzieren, die von den Developern in den USA ohnehin zu erbringen sind. In der Regel errichten vor allem Developer von Einfamilienhausgebieten die entsprechende Infrastruktur, insbesondere auch das Abwassersystem selbst. Development Inmpact Fees werden vor Erteilung der Planungsgenehmigung erhoben. Die Impact Fees werden meist als ein Dollarbetrag pro Quadratmeter oder pro Einfamilienhaus erhoben. Die Abgaben können im Einzelfall bis zu 3.000 Dollar pro Haus erreichen. Diese Abgaben sind nicht mit Erschließungsbeiträgen zu verwechseln, wie sie in der Bundesrepublik erhoben werden. Impact Fees sind aber auch keine Steuer, die nach allgemein politischen Gesichtspunkten festgesetzt wird. Der Äquivalenzgedanke steht bei dieser Erhebung im Vordergrund.

Zum Inhalt der Impact Fees[8]

Impact Fees weden als pauschale Gebühr zu einem Quasipreis für Entwicklungsrechte. Sie müssen allerdings in einem rationalen Nexus zu den Investitionen stehen. Sie werden in der Regel aufgrund von Ermächtigungen durch die staatliche Gesetzgebung der einzelnen amerikanischen Staaten erhoben. Mit rationalem Nexus werden folgende Komponenten umfaßt:

– Notwendigkeit: Die erhebende Behörde oder Gemeinde muß zeigen, daß sie selbst als Folge der neu zugelassenen Investitionen eigene Folgemaßnahmen durchführen muß.
– Proportionalität: Es muß ein rationaler, quantitativer Zusammenhang zwischen Ausgaben und den Vorteilen bestehen.
– Berechenbarkeit: Die Impact Fees müssen für Zwecke ausgegeben werden, für die sie erhoben wurden.
– Kongruenz: Die Impact Fees müssen für Zwecke erhoben werden, die kongruent zu den belasteten Entwicklungsmaßnahmen sind. So wurde z.B. entschieden, daß Impact Fees nicht für den Bau von Schulen verwendet werden können, nachdem sie beim Bau von Wohnungen für alte Menschen erhoben wurden. Ähnlich wie bei den Linkage-Strategien muß ein sachlicher Zusammenhang bestehen, d.h. für Wohngebiete können Parks oder andere, den Bewohner dienenden Leistungen finanziert werden. Dabei muß der Park nicht in unmittelbarer Nähe zu den belasteten neuen Wohngebieten liegen. Als fraglich wurde es dagegen bezeichnet, ob Impact Fees auf eine Büroentwicklung zur Schaffung von Parkanlagen herangezogen werden können.

Die lockeren Definitionen erlauben es, sehr weitgehende Folgeleistungen aus Impact Fees zu finanzieren. Im Ergebnis kann man sie als sehr flexible Folgekostenbeisträge bezeichnen. Hintergrund ist auch, daß die Bürger vieler Städte immer weniger bereit sind, Steuererhöhungen für die Wachstumskosten neuer Siedlungen hinzunehmen oder überhaupt hohe Ausgaben zugunsten neuer Siedlungen zu tragen. D.h., die Neuentwicklung muß für die Städte weitgehend kostenneutral sein. Im Ergebnis läuft dies darauf hinaus, daß die Entwicklungen von Siedlungen nicht mehr als eine Art Gemeinwohlleistung der Kommune angesehen wird. Die Kommune betrachtet sich selbst als eine öffentliche Entwicklungsgesellschaft, die Folgekosten der Entwicklung aus Gebühren und Belastungen finanziert, die den Investoren und Eigentümern in den neuen Siedlungen angelastet werden. Anstelle des engen Planungs-Wertausgleichdenkens, bei dem die öffentliche Hand einen Vorteil abschöpft, der durch ihre Planungsleistungen entsteht und bei dem versucht wird die Gleichheit zwischen unterschiedlichen Grundstückseigentümern herzustellen, geht es bei Planning Gain und Impact Fee – Denken darum, daß die Kommune eine Gesamtentwicklung vornimmt die sich selbst tragen soll. Die Kommune wird zu einer Art Stadtunternehmer. Es wird versucht, den Bereich des Äquivalenzdenkens, in dem Leistung und Gegenleistung im Vordergrund stehen, allmählich auszuweiten. Es kommt zu einem «Marktdenken» bei der Bereitstellung öffentlicher Infrastrukturleistungen. Dieser Gedanke ist in der amerikanischen Diskussion bisher klarer artikuliert als in der englischen Diskussion.

Hinweis: Das Äquivalenzdenken ist in den USA generell sehr viel weiter ausgebildet als bei uns. So werden z.B. Kongreßzentren aus Steuern auf Hotelbetten finanziert, da die Hotels die Hauptnutznießer solcher Zentren sind und Kongreßzentren in der Regel nur mit Defiziten betrieben werden können. In der Bundesrepublik oder in Großbritannien werden solche Leistungen noch aus allgemeinen kommunalen Einnahmen finanziert.

Bewertung

Aus einer theoretischen Perspektive kann man eine solche Entwicklung begrüßen, da solche Zurechnungen langfristig zu einer besseren Allokation von Ressourcen führen. Gegenwärtig werden Planungsrechte preisfrei zur Verfügung gestellt. Infrastruktur und Folgelasten werden von der Allgemeinheit getragen. D.h., die Entwicklung neuer Siedlungen erscheint dem einzelnen Developer als viel zu preisgünstig.

Als neues Problem taucht bei der Erhebung von Impact Fees auf, daß Neuentwicklungen sehr viel weiter definierte Folgekosten tragen müssen. Grundstücksentwicklungen werden verteuert mt der Folge, daß Alteigentümer Vorteile erzielen, weil das gesamte Immobilienpreisniveau bzw. die Nutzungspreise von Immobilien durch eine umfassendere Zurechnung der Folgekosten angehoben wird. Konsequenterweise müßten diese, durch Wettbewerb nicht mehr abbaubaren Renten, durch hohe steuerliche Belastungen z.T. abgeschöpft werden. Für sich genommen führen Impact Fees zu einer besseren Zurechnung und Verteilung der Folgekosten bei neuen Entwicklungen. Das Verursacherprinzip wird in einem weiteren Sinn erreicht als durch die Erschließungsbeiträge. Eine solche breite Anwendung des Äquivalenzdenkens dürfte langfristig konsequent sein, wenn man berücksichtigt, daß z.B. aufgrund ökologisch schärfer werdender Begrenzungen Baurechte auf Dauer knapp gehalten werden. Das Knapphalten der Baurechte als Rationierungsstrategie führt jedoch zu unerträglichen Verteilungswirkungen, weil die planungsbedingten Wertsteigerungen zunehmen. Sehen sich die Kommunen dagegen stärker als Stadtentwicklungsunternehmen, die Erweiterungen nur noch vornehmen, wenn diese sich umfassend selbst finanzieren, dann erhält die Verknappung von Baurechten eine zusätzliche Begründung. Die negativen Verteilungswirkungen von planerischen Rationierungsentscheidungen werden abgebaut.

Um Impact Fees zu begründen, müßten die Gemeinden sehr viel detailliertere Buchhaltungssysteme entwickeln, aus denen deutlich wird, welche Zusammenhänge zwischen Siedlungsentwicklungen und bestimmten Folgelasten besteht. Es müßten dann spezielle Finanzierungskreisläufe entwickelt werden, d.h. die einzelne Siedlung wird zu einer Kostenstelle der die an verschiedenen Stellen des Systems anfallenden Kosten zugerechnet werden.

Die Neubausiedlung oder ein großes Investitionsprojekt würden für die Kommune zu einem «Profitcenter», das «seine Kosten» selbst tragen muß.

Diese theoretischen Implikationen sind bisher im Zusammenhang mit Planning Agreements und Impacts Fees nicht ausdiskutiert und z.T. auch nicht artikuliert. Auf jeden Fall entsteht durch die Planning Agreements, in denen versucht wird Planning Gain für die öffentliche Hand zu mobilisieren, ein neues Verständnis der Rolle der Kommunen bei der Entwicklung von Grundstücken. Die Vergabe von Baurechten gilt weniger als hoheitlicher Akt, bei dem aus dem Gemeinwohl abgeleitete Rechte gewährt werden. Die Gemeinde wird vielmehr zu einem Mitentwickler eines Gebiets oder bei der Realisierung eines speziellen Vorhabens. Die Entwicklung von Immobilienprojekten wird auch aus der Sicht der Kommune zu einem wirtschaftlich zu kalkulierenden

Development-Vorgang, bei dem sie selbst durch Impact Fees die Preise für die Teile der Entwicklung festlegt, die nicht von den privaten Investoren getragen wird.

Die Rolle der Banken

Hohe Risikobereitschaft

In der Vergangenheit wurden Immobilieninvestitionen weitgehend von den institutionellen Anlegern finanziert und auch bewirtschaftet (Versicherungen, Pension Funds). Sie haben seit Ende der 80er Jahre ihr Engagement am Immobilienmarkt zumindest relativ reduziert, d.h. sie haben das kräftige Wachstum nicht mitfinanziert, weil nach ihrer Auffassung der Markt zum Teil übersättigt war oder weil andere Anlagemöglichkeiten höhere Renditen ermöglichten. Damit haben die klassischen Eigenkapitalinvestoren bei den Neuanlagen ihre Marktanteile verringert. Der Boom wurde im wesentlichen zu einer Leistung der Property Companies und damit zu einer Finanzierungsaufgabe für die Banken. Als besonderes Merkmal kam hinzu, daß zum Teil kleine, junge Property Companies riesige Projekte realisierten. Als herausragendes Beispiel sind die Projekte von Stanhope zu erwähnen. Als kleine Gesellschaft hat Stanhope Mitte der 80er Jahre in Broadgate einen Komplex mit rd. 300 000 qm Büroflächen entwickelt (vgl. Fallstudie in Teil B). Solche hohen Wachstumsraten und die dafür erforderlichen Finanzierungen wären nicht möglich gwesen, wenn die Banken die Immobilienfinanzierung nicht als neue Wachstumsbranche «entdeckt» hätten. Dabei haben sich die Banken sehr viel risikofreudiger als in der Bundesrepublik verhalten. Während die Hypothekenbanken in der Bundesrepublik durch rechtliche Rahmenbedingungen zur «Risikoscheu» verpflichtet sind, können englische Banken in der Beleihung sehr viel stärker nach eigenen Einschätzungen und nach ihrer eigenen Risikobereitschaft vorgehen. Außerdem spielt eine Rolle, daß der Wettbewerb zwischen Banken auf dem internationalen Finanzmarkt in London sehr viel schärfer ist als in der Bundesrepublik. Durch diesen Wettbewerb wurden Banken «gezwungen», höhere Risiken einzugehen. Nur so ist zu erklären, daß es im Verlauf der 80er Jahre immer häufiger vorkam, daß Banken sogenannte Non Recourse Loans gewährten, d.h. Darlehen, die nur am Objekt abgesichert werden, ohne Rückgriff auf das Vermögen der Firma. Die Banken werden damit de facto sehr viel stärker als bei uns zu Mitentwicklern. Dies äußert sich z.T. auch darin, daß sie am Gewinn der Projekte beteiligt werden.

Unterschiede zur Bundesrepublik

Die Welle der Finanzierungen nach 1985 in Erwartung des Big Bang kam zum Teil von internationalen Banken mit geringer oder keiner Erfahrung am Property Markt. Dies hat mit dazu beigetragen, daß der exorbitante Boom überhaupt finanzierbar wurde. In der Bundesrepublik erscheinen solche Entwicklungen kaum möglich. Dabei spielt das Zweiebenensystem eine besondere Rolle. Hypothekenbanken dürfen nur bis zu 60% des Verkehrswertes beleihen. Spitzenfinanzierungen müssen damit von anderen Banken bereitgestellt werden. Dies führt automatisch zu einer hohen Risikoempfindlichkeit. Wer lediglich die letzten 20-30% finanziert, muß damit rechnen, daß im Falle einer Krise des Projektes sein gesamtes Darlehen in Frage steht. Wer dagegen 80-90% aus einer Hand bereitstellt, kann damit rechnen, allenfalls eine kleine Quote seines Kredits zu verlieren. Das Zweiebenensystem in der Bundesrepublik führt zu größerer Risikoscheu und größerer Vorsicht. Dies erklärt zum Teil, warum die Investitionszyklen und Preiszyklen in der Bundesrepublik weniger ausgeprägt sind.

Zur Refinanzierung

Wie erwähnt werden in Großbritannien auch Immobilienfinanzierungen mit sehr kurzfristigen Refinanzierungsmitteln vorgenommen. Ein Pfandbriefmarkt besteht de facto nicht. Allerdings werden im Einzelfall zur Bewältigung großer Finanzierungsvolumen sehr kreative und flexible Lösungen gefunden. So kann die Neuplazierung von Aktien mit Darlehenfinanzierungen kombiniert werden. Gleichzeitig sind gerade am Immobilienmarkt «Securitization-Techniken» in großem Stil zur Anwendung gekommen. Hierbei werden die Forderungen an Kreditnehmer an andere Finanzierungsinstitutionen abgetreten. Dabei werden größere Beträge mit unterschiedlicher Zusammensetzung und Stückelung gebündelt und als solche auch prozentweise an Anleger verkauft. Durch Versicherungsverträge werden die Erträge dieser neugeschaffenen Wertpapiere in ihrer Rendite z.T. abgesichert. In der Bundesrepublik sind solche Techniken bisher nicht praktiziert worden. Dies geht im wesentlichen darauf zurück, daß mit den Pfandbriefen und den Refinanzierungstechniken der Hypothekenbanken bisher alle Bedürfnisse abgedeckt werden konnten. Ausländische Experten vertreten immer wieder die Position, daß sich auch in der Bundesrepublik langfristig flexible Techniken der Refinanzierung durchsetzen dürften. Sie verweisen darauf, daß die Tage des deutschen Pfandbriefes allein deshalb gezählt seien, weil der Ausschluß von Tilgungsmöglichkeiten von Finanzierungen zu erheblichen Starrheiten auf einem flexiblen Markt mit hohen Transaktionsvolumen führen wird. Dem steht entgegen, daß die deutschen Finanzierungstechniken bisher ohne Mühe die erforderlichen Volumen mobilisieren konnten. Deutsche Hypothekenbanken verweisen im Gegenteil darauf, daß ihre Finanzierungen im Ausland sehr gefragt sind. So werden mehr und mehr ausländische Immobilien mit DM-Hypotheken finanziert. Dabei kommt den deutschen Banken natürlich zugute, daß mit einm Zinssatz von rd. 9% bei Zinsen in Großbritannien von etwa 14% selbst Festzinshypotheken mit einer Laufzeit von 5 Jahren sehr attraktiv sind, da kaum zu erwarten ist, daß die Zinsen in Großbritannien sehr rasch auf das deutsche Niveau absinken werden.

Man kann deshalb davon ausgehen, daß vom Kapitalmarkt her nur allmähliche, strukturelle Einflüsse auf den Immobilienmarkt einwirken werden. Es ist anzunehmen, daß der wachsende Anteil der ausländischen Investoren in einigen Großstädten den Wettbewerb verschärft, und daß aufgrund risikofreudigerer Finanzierungen auch ein Element größerer zyklischer Instabilität möglich wird.

6
Die Rolle der Entwicklungsgesellschaften und der Property Companies

Die innere Stadtentwicklung Londons war in den letzten Jahren durch eine Serie großer komplexer Büroprojekte gekennzeichnet: Broadgate, Canary Wharf, London Bridge, die Bebaung von Kings Cross und zahlreiche andere Projekte erreichen jeweils Größendimensionen, die in der Vergangenheit kaum jemals zu beobachten waren. Ohne das dies im Einzelfall nachgewiesen oder überprüft werden kann, entstand in den Interviews der Eindruck, daß private Unternehmen bei der Vorbereitung und Entwicklung dieser Projekte eine sehr viel aktivere Rolle spielten als sie in der Bundesrepublik üblich ist.

Zwar kann man seit einiger Zeit auch in Frankfurt, Hamburg oder in jüngster Zeit in Berlin feststellen, daß Investoren mit sehr detaillierten Vorstellungen und Vorplanungen an die planende Verwaltung herantreten, um für größere Gebiete oder Grundstücke Investitionsmaßnahmen anzuregen. Praktisch alle Hochhausentscheidungen in Frankfurt waren Folge solcher intensiven Vorbereitungen durch private Investoren. Das jüngste Projekt am Hemmerichsweg in Frankfurt wurde durch ein Standortgutachten, das vom Investor in Auftrag gegeben wurde, vorbereitet. Es wurden detaillierte Analysen über die Auswirkungen der Hochhausbebauung auf die Nachbarschaft vorgelegt. Es wurde geprüft, inwieweit alternative Nutzungen z.B. eine Wohnbebauung als Ergänzung sinnvoll und möglich erscheint. Gestützt auf diese Vorklärungen wurde ein internationales Wettbewerbsverfahren finanziert. Die Stadt hat diese Vorbereitungen mit Wohlwollen begleitet und steht jetzt vor der Frage, ob sie die entsprechenden Planungen vornehmen will, um die vorgeschlagenen Projekte zu realisieren.

Verfahren dieser Art sind in Großbritannien im Laufe der 80er Jahre immer üblicher geworden. Dabei hat die Planung zu einer aktiveren Rolle der privaten Investoren dadurch eingeladen, daß sie in der Mitte der 60er Jahre in der City die zulässigen Baudichten generell erhöhte. Als sehr wichtig hat sich auch herausgestellt, daß in den Baunutzungskategorien die Sonderkategorie «Light Industries» aufgehoben wurde und gleichgesetzt wurde mit Bürobebauung. Dies hat z.B. am Kings Cross die Vorschläge zu einer Bürobebauung zur Folge gehabt. Mit dieser Verschmelzung von bisher spezialisierteren Kategorien hat der Gesetzgeber im übrigen der Tatsache Rechnung getragen, daß immer mehr «Quasibüros» entstehen, d.h. gewerbliche Gebäude und Anlagen, die in ihren Bauformen und ihren Auswirkungen auf die Städte kaum von reinen Bürokomplexen zu unterscheiden sind (geringer Gütertransport, keine Lärm- und Schmutzimmissionen, hoher Anteil von Bürobeschäftigen).

Gestützt auf sehr großzügige Rahmenbedingungen haben bei wachsender Nachfrage die privaten Entwicklungsgesellschaften die entstehenden Spielräume ausgenutzt und eigene, sehr detaillierte Nutzung- und Investitionskonzepte für große Gebiete vorgelegt. Dabei

233

hat sich gezeigt, daß aufgrund der steigenden Anforderungen an die Architektur, an das Image von Gebäuden sowie an die Qualität von Arbeitsplätzen die Rolle der Planung als eine Mindestsicherung von Qualität immer weniger wichtig wird. So wurden die Vorschriften über die innere Gestaltung von Gebäuden generell aufgelockert, mit dem Ergebnis, daß die Investoren eine höhere Wahlfreiheit haben. Sie können Grundstücke mit sehr tiefen, kompakten Gebäuden hochverdichtet nutzen oder nach wie vor Hochhäuser erricten. Diese Wahlfreiheit hat zu interessanten Vergleichsmöglichkeiten unterschiedlicher Bebauungskonzepte geführt. In der Sprache traditioneller Planungsdiskussionen in der Bundesrepublik ist die Planung in London sehr viel stärker als in der Vergangenheit Anpassungsplanung geworden. Dies begünstigt eine große Bandbreite von Bauformen. Als Gegenpole können z.B. die Investitionen von Broadgate (Stanhope) und Canary Wharf (Oympia und York) gelten. Gleichzeitig hat die Vielfalt der Möglichkeiten auch zu einem sehr viel heterogeneren Erscheinungsbild insbesondere der Londoner City geführt. Dies wird z.b. in einer schon wirr zu nennenden Vielfalt der Bauformen in der neuen Skyline der City sichtbar. Es ist fraglich, ob ein so stark zurückgenommenes Planungsverständnis auf Dauer die optimale Lösung darstellt. So wäre es im Kontrast durchaus möglich gewesen, durch Design Guidelines die Vielfalt der Ausdrucks- und Erscheinungsformen zu reduzieren und eine Entwicklung der City und ihrer Randbereiche herbeizuführen, die durch eine größe Ruhe und Gleichmäßigkeit der Bau- und Ausdrucksformen gekennzeichnet gewesen wäre.

Den Extremfall einer solchen «Market-Led Development» begegnet man in den Docklands, wo die LDDC bewußt ihre im Prinzip vorhandenen Steuerungsmöglichkeiten nicht benutzt hat, um auf das Erscheinungsbild und die Gestaltung der neubebauten Gebiete einzuwirken. Sie hat sich mehr als ein nachgiebiger und toleranter Koordinator betätigt, der im wesentlichen darauf achtet, die Funktionsfähigkeit der benachbarten Baukomplexe zu sichern, soweit es bei dem extremen Bautempo überhaupt möglich war.

Für die Bundesrepublik empfiehlt es sich, die weitere Entwicklung sehr sorgfältig zu beobachten, um daraus Konsequenzen für das eigene Planungsverständis zu ziehen. Frankfurt ist in der Bauentwicklung der jüngsten Vergangenheit der Londoner Entwicklung noch am nächsten gekommen. Dies gilt insbesondere auch, was die Bauformen und das Erscheinungsbild angeht. In Hamburg stehen mit der Umwidmung der Speicherstadt und der angrenzenden Gebiete ähnliche Entscheidungen an. In Berlin liegt auf der Hand, daß durch die rasche Bebauung großer, bisher brachliegender Gebiete auch die Frage entschieden werden muß, inwieweit die privaten Investoren über die Gestalt und das Erscheinungsbild der Stadt entscheiden oder inwieweit Planung hier koordinierend und eingrenzend wirken will.

In jedem Fall kann man auch in der Bundesrepublik davon ausgehen, daß große innerstädtische Investitionsprojekte sehr weitgehend von privaten Investoren vorbereitet und initiiert werden. Damit ist allerdings nicht gesagt, daß sich die Planung in ihren Dichte- und Bauformenvorgaben an die Vorstellungen der Investoren anpassen muß. In einer partnerschaftlichen Entwicklung ist durchaus vorstellbar, daß größere Areale in enger Absprache und Koordinierung auch der architektonischen Ausdrucks- und Gestaltungsformen entwickelt werden. Fragen der Designkoordinierung werden bisher kaum diskutiert. Angesichts des vorherrschenden Eklektizismus in der Architektur und des Bürobaubooms, der sich in mehreren Innenstädten der Bundesrepublik abzeichnet, scheint eine solche Designkoordinierung jedoch nahezuliegen.

7

Folgerungen und Fazit

Optimale Rahmenbedingungen schaffen

Die Folgerungen und Empfehlungen zielen darauf ab, gestützt auf die Erfahrungen am Immobilienmarkt Großbritanniens insbesondere des Marktes in London, Rahmenbedingungen bzw. Instrumente vorzuschlagen, die zu optimalen Ergebnissen führen. Dabei liegt auf der Hand, daß die Ziele, die bei der Entwicklung des Immobilienmarktes in Zukunft erreicht werden sollen, im Vergleich zur Vergangenheit sehr viel komplexer geworden sind.

– Angesichts steigender Nachfrage insbesondere im Bürobereich und steigender Nachfrage nach Flächen für gewerbliche Investitionen, muß jedes System elastisch reagieren, d.h. die gegenwärtig zum Teil beobachtbaren Starrheiten auf der Angebotsseite, die zurückgehen auf enge planerische Vorgaben, sollten im Hinblick auf eine nachfrageorientierte Investitionstätigkeit aufgelockert werden.

– Gleichzeitig soll rreicht werden, daß sich das Erscheinungsbild der Städte verbessert und ihre Attraktivität steigt.

– Der Leistungswettbewerb der Anbieter und Nachfrager soll nicht behindert werden.

– Angesichts der wachsenden Bedeutung öffentlicher Vorleistungen für alle privaten Investitionen wird das Thema der Zurechnung von öffentlichen Leistungen, bzw. der Berücksichtigung von Folgekosten die private Investitionen hervorrufen, immer dringlicher. Es ist nicht einmal in Ansätzen befriedigend gelöst.

– Die wichtiger gewordenen ökologischen Ziele machen es immer weniger sinnvoll über Verbilligung von Bauland, Gewerbepolitik und Ansiedlungspolitik zu betreiben. Immobilieninvestitionen sollten im Gegenteil möglichst alle Folgekosten tragen. Grundstückspreise sollten den tatsächlichen Knappheiten entsprechen.

– Als instrumentales Ziel ist eine hohe Markttransparenz, eine hohe Qualität der Marktorganisation einschließlich der den Immobilienmarkt organisierenden Dienstleistungen notwendig.

Damit kann man die Zielsetzungen oder Aufgaben unter denen die Empfehlungen stehen mit wenigen Schlagworten zusammenfassen: hohe Qualität der Investitionen, scharfer Leistungswettbewerb, Markttransparenz, hohe Qualität der Dienstleistungen, die zur Funktionsfähigkeit der Märkte erforderlich sind, enge partnerschaftliche Kooperation zwischen planenden Verwaltungen und privaten Investoren, hohe Qualität des Erscheinungsbildes und der Gestaltung der Städte bzw. ihres Umlandes, Zurechnung von Folgekosten und hohe Markttransparenz.

Es liegt auf der Hand, daß diese Anforderungen und Zielsetzungen auf den

Immobilienmärkten Großbritanniens bei weitem nicht optimal erfüllt sind. In Teilbereichen sind im Vergleich zur Bundesrepublik jedoch Vorteile zu erkennen. Hier können Nachahmungen oder analoge Regelungen vorgeschlagen werden. Unabhängig davon bleibt die Aufgabe durch ständige Beobachtungen der Entwicklung der Märkte in den Großstädten der Bundesrepublik Verbesserungen der Rahmenbedingungen und der eigenen Märkte zu erreichen bzw. die Rahmenbedingungen so zu verändern, daß private Investoren und öffentliche Planung möglichst produktiv miteinander kooperieren.

Marktorganisation–Markttransparenz

Markttransparenz verbessern
Hochentwickelte Märkte mit einem hohen Transaktionsvolumen und einem hohen Anteil professioneller Anleger, die ihre Vermögensportfolios ständig optimieren wollen, benötigen eine hohe Markttransparenz durch leicht zugängliche Informationen über:
- Investitionsplanungen und im Bau befindliche Projekte einschließlich ihrer Realisierungsfristen.
- Informationen über Mieten und Renditen, Kostenstrukturen und Transaktionsvolumen.

Folgende Maßnahmen können dieses Ziel fördern:
- Aufhebung des Datenschutz für Bauanträge. Gebäudeinvestitionen haben nicht nur eine private Komponente. Alle Gebäude sind auch öffentliche Gebäude, weil sie Teile gemeinsamer Umwelt und gemeinsamer Märkte sind. Bauanträge sollten dem englischen Beispiel folgend deshalb von dem Zeitpunkt an, in dem sie förmlich gestellt sind, als öffentlich zugängliche Information behandelt werden. Durch solche erhöhte Transparenz können natürlich auch Widerstände gegen Projekte mobilisiert werden, einen Effekt den man akzeptieren muß. Für die Investoren ist jedoch wichtig, daß sie auf diese Weise einen Überblick über Investitionsplanungen in der näheren und weiteren Umgebung der geplanten Objekte erhalten. Dies kann die Sicherheit der Planungen erhöhen und wirkt langfristig risikomindernd.
- Die Gutachterausschußdaten enthalten wertvolles Material über Umsätze von Grundstücken und zum Teil auch über die Mieten vor der Veräußerung. Diese Daten werden bisher nur unzureichend ausgewertet. Vor allem die Informationen über gewerbliche Transaktionen verrotten weitgehend ungenutzt bei den Kommunen. Hier wäre in einem Pilotprojekt zu prüfen, wie in Zusammenarbeit mit den Kommunen und der örtlichen Immobilienwirtschaft ein für die Investoren wirklich nützliches Marktinformationssyste aufgebaut werden kann. Der Immobilienmarkt in Großbritannien demonstriert, es gibt einen wachsenden Informationsbedarf an Umsatz- und Renditedaten für gewerbliche Investoren. Diese Informationen können auch privatwirtschaftlich aufbereitet und bereitgestellt werden. Die Auswertungen der Daten der Gutachterausschüsse bleiben weit hinter dem zurück, was mit vertretbarem Aufwand möglich erscheint.

Erhöhung der Fungibilität von Immobilienanlagen
Eine wirtschaftliche Ausnutzung der Immobilienbestände setzt eine hohe Fungibilität der Objekte voraus. In der Tendenz sollten Rahmenbedingungen entstehen, die Immobilienanlagen in ihrer Rentabilität möglichst genau so leicht und rasch verwertbar machen, wie Wertpapieranlagen. Die Transaktionskosten wären möglichst niedrig zu halten.

Diese Voraussetzungen hängen von verschiedenen Faktoren ab:
- Professionelle Dienste: Angesichts des im Prinzip inhomogenen Immobilienmarktes, seiner komplexen Kostenstruktur und den im Einzelfall uneinheitlichen Verwertungsverträgen, sind professionelle Dienste eine wichtige Voraussetzung für leistungsfähige Märkte.

Das englische Surveyor System mit seinem hohen fachlichen Niveau und den einheitlichen Standards bei der Bewertung, beim Management oder bei der Informationsaufbereitung erleichtert die Marktbeziehungen erheblich. In der Bundesrepublik bestehen keine entsprechenden spezialisierten Fachausbildungen. Die European Business Scool bietet gegenwärtig Kurse für Immobilienspezialisten an. Einzelne Lehrstühle (Siegen) befassen sich mit dem Thema. In Dortmund gibt es einen Schwerpunkt für Raumplanung. Die Ausbildung ist jedoch stark auf den öffentlichen Sektor hin orientiert. Eine entsprechende Ausrichtung auf die Bedürfnisse der privaten Immobilienwirtschaft besteht nicht. Diese Lücke sollte systematisch durch entsprechende Ausbildungsangebote geschlossen werden. Darüber hinaus wäre zu prüfen, ob nicht wie z.B. an amerikanischen Universitäten üblich, Sommerkurse und Fortbildungskurse systematisch an Hochschulen oder Fachhochschulen angeboten werden können.
- Senkung der Transaktionskosten: Zu einer Senkung der Transaktionskosten wird es durch steigende Umsätze kommen. Darüber hinaus wäre dem Markt doch gedient, wenn Makler als Berater einer Partei auftreten und ihre Rolle weniger transaktions- und mehr bewirtschaftungsorientiert sehen würden. Es ist zu prüfen, ob hier entsprechende Regulierungen möglich sind, die in diese Richtung wirken.

Empfehlungen für den öffentlichen Sektor

Verwertungspraxis von Grundstücken

Es zeichnet sich ab, daß aufgrund der ökologischen Rahmenbedingungen Baurechte und Bauland bzw. Infrastruktur auf Dauer relativ zu anderen Gütern immer knapper werden. Dies muß weitreichende Konsequenzen für das Steuersystem haben. Darüber hinaus ist die Grundstückswirtschaft in mehrfacher Hinsicht betroffen.

Verkaufspraxis

Gegenwärtig werden Grundstücke von Kommunen fast regelmäßig zum Verkehrswert veräußert wenn sie für private Zwecke benötigt werden. Da die Kommunen auf Dauer Flexibilitätsreserven und Eingriffsmöglichkeiten am Bodenmarkt benötigen, sollten sie Grundstücke insbesondere dort, wo es sich um hochwertige Lagen handelt nicht mehr veräußern, sondern entweder Erbbaurechtsverträge vergeben, oder aber Joint Venture-Lösungen angehen. Wie das Beispiel Großbritannien zeigt, gewöhnen sich Banken an eine solche Praxis sehr rasch, wenn sie generell üblich wird. Die Beleihungsvorbehalte sind weitgehend Scheinargumente.

Die Kommunen selbst sollten ihre eigenen Immobilienbestände in selbständige Immobilienunternehmen ausgründen. Erste Tendenzen hierfür gibt es auch in Großbritannien. Solche Ausgründungen und verselbständigte Grundstücksorganisationen gibt es allerdings bisher überwiegend nur im privaten Sektor. Im kommunalen Bereich beginnen gegenwärtig einzelne Boroughs in London ihre Immobilieninteressen in eigenen Organisationen zu verselbständigen (Borough of Lambers). In einem hochentwickelten Stadium müßten die einzelnen Organisationseinheiten der ommunal-verwaltung oder von großen Unternehmen von den ausgegründeten Immobiliengesell-schaften Büros und andere Flächen zu Marktpreisen anmieten. Auf diese Weise würde sichtbar, wie viele öffentliche Stellen wertvolle Grundstücke nutzen, die angesichts der geringen Kundenkontakte nicht erforderlich sind.

Behandlung von Planungsgewinnen Durch die Entwicklungsmaßnahmen nach 165 ff Baugesetzbuch wurden Kommunen wieder in die Lage versetzt, bei größeren Erschließungen, planungsbedingte Wertsteigerungen zur Finanzierung der Infrastruktur heranzuziehen. Unabhängig von solchen geschlossenen Maßnahmen entstehen Sonder-vorteile bei zahlreichen anderen Planungen. Angesicht der knapper werdenden Baurechte nimmt der vermögenswerte Vorteil durch die Gewährung von Baurechten ständig zu. Hier sollte systematisch geprüft werden, ob Lösungen zu finden sind. Die angelsächsische Diskussion ist auch hier noch nicht zu einem abschließenden Ergebnis gekommen. Die Diskussion zeigt jedoch, daß unter den veränderten Rahmenbeding-ungen Themen dieser Art in Zukunft gewichtiger werden dürften. Als Lösungen kommen in Frage:

- Steuerrechtliche Lösungen (zeitnahe Grundstücksbewertung und höhere Besteuerung im Rahmen einer inhaltlich veränderten Grundsteuer)
- Abgabenlösungen zur Abgeltung von Ausstrahlungsvorteilen großer kommunaler Investitionen (U-Bahnhöfe, Flughäfen etc.)
- Entwicklungssteuer bei der Erschließung neuen Baulands. Einer solchen Ent-wicklungssteuer könnte über die Erschließungsbeiträge hinausgehen und wirtschaftlich als ein Preis für die Gewährung von Bebauungsmöglichkeiten betrachtet werden. Durch eine Entwicklungssteuer würden auch Folgelasten, die nicht unmittelbar mit der Grundstückserschließung zusammenhängen (Ausweitung des Schulsystems, allge-meine Verwaltung etc.) abgedeckt werden.

Planungskonzepte zur Einzelhandelsentwicklung Die wenigen großen Einzelhandels- und Freizeitzentren, die in Großbritannien entstanden sind werden gegenwärtig in ihren Auswirkungen auf die angrenzenden Städte und Gemeinden untersucht. Dahinter steht eine kritische Diskussion, die der restriktiven Planungspraxis bei der Ausweisung von großen Einkaufszentren im Umland von Agglomerationen kritisch gegenübersteht. Wie die Fachdiskussion bei der Bewertung entsprechender Projekte im Ruhrgebiet gezeigt hat, fehlte es damals an konkreten europäischen Ergebnissen. Es empfiehlt sich, daß Ergebnis der Diskussion in Großbritannien in seinen Folgerungen für die Bundes-republik genauer auszuwerten.

Leistungsfähige Entwicklungsgesellschaften In der Bundesrepublik haben sich auf der Landesebene Entwicklungsgesellschaften nach dem Ressortprinzip gebildet, d.h. es gibt ein Nebeneinander von Wirtschaftsförderungsgesellschaften und Landesentwicklungs-gesellschaften, die stärker Grundstücksbezogen agieren. Mit der Welsh Development Agency und der Scottish Devolopment Agency gibt es in Großbritannien Beispiele von Entwicklungsorganisationen, die gleichzeitig Grundstücksentwicklung betreiben und Wirtschaftsförderungsaufgaben übernehmen. Die Urban Development Corporations, an der Spitze die LDDC sind stärker Grundstücksorientiert, haben darüber hinaus aber auch Wirtschaftsförderungsaufgaben insbesondere im Bereich der Akquisition von Investoren erfüllen. Es spricht vieles dafür, die aus dem Ressortdenken entstandene Arbeits-aufteilung in der Bundesrepublik zu überwinden. Dies gilt insbesondere im Hinblick auf Entwicklungsorganisationen, die jetzt in Ostdeutschland aufgebaut werden müssen. Hier könnten die Erfahrungen integrierter Organisationen in Großbritannien systematisch ausgewertet werden.

Deregulierung bei den Vorschriften zum Bau von Büros Wie am Beispiel verschied-ener Büroprojekte in Großbritannien deutlich wird, konkurrieren dort auf der einen Seite Hochhauslösungen (Skyscraper) mit kompakten niedriggeschossigen Lösungen (Ground-scraper), wobei die niedriggeschossigen Lösungen – 6 bis 9 Stockwerke – in der Regel hohe Dichten durch tiefe Gebäude erreichen. Diese tiefen Gebäue erfordern häufig Großraumbüros oder Bauformen, bei denen kleine Innenhöfe, Lichthöfe und ähnliches vorgesehen sind.

In der Bundesrepublik sind solche Lösungen zum Teil aufgrund von Arbeitsschutz-vorschriften (Mindestentfernung der Arbeitsplätze zum nächsten Fenster) nicht möglich. Wie das Beispiel Großbritannien zeigt, werden unterschiedliche Bauformen am Markt auch von hochaqualifizierten Arbeitskräften akzeptiert. Hier sollte geprüft werden, ob durch Deregulierungen nicht eine größere Bandbreite von Bauformen auch in der Bundesrepublik zugelassen wird, weil die historisch gewachsenen Arbeitsschutz-vorschriften z. T. obsolet geworden sind.

Verstärkung selbständiger Pensionskassen Die Praxis deutscher Großunternehmen, Pansionverpflichtungen als eigene steuerbegünstigte Rücklage in der Billanz einzustellen, erschwert die Entfaltung des Kapitalmarktes einschließlich leistungsfähiger Immobilienfonds. Der Leistungswettbewerbs wird um knappes Kapital verringert. Die englische Lösungen bei der Property Companies oder Property Unit Trusts oder ver-schiedene Investmentgesellschaften um die Pensionsrückstellung konkurrieren ermöglicht mehr Wettbewerb und stärkt professionelle Marktorganisationen. Entwicklung in diese Richtungen setzen jedoch grundlegende Änderungen bei der Behandlung von Pensions-rückstellungen voraus, die weit über die Fragen der Immobilieninvestitionen hinausgehen.

Fazit

Die Unterschiede der Funktionsweise von Immobilienmärkten zwischen Großbritannien und der Bundesrepublik beruhen auf verschiedenen Faktoren:
- Unterschiedliche steuerliche Regelungen (steuerbefreite Wertsteigerungen für private Investoren in der Bundesrepublik, offener Wettbewerb um Pensionsrückstellungen in Großbritannien...).
- Auf Unterschieden der Berufs- und Ausbildungstraditionen (Surveyor, verschiedene Dienstleistungsberufe...).
- Auf unterschiedlichen Rechtstraditionen (hohe Verbreitung von erbbaurechtsähnlichen Lösungen in Großbritannien).
- Auf hochentwickelten Informationssystemen, die allerdings z.T. fast automatische Folge eines hochentwickelten und differenzierten Marktes sind.

Die Unterschiede des Planungsrechts haben, wenn man von der Sondersituation in den Gebieten der Urban Development Corporations absieht keine entscheidenden Auswirkungen. Allerdings hat sich in den 80er Jahren eine liberalere Handhabung der Planungsrechte durchgesetzt, mit der Folge, daß private Investoren leichter als in der Bundesrepublik Großprojekte in eigener Regie planen und entwickeln können.

Mehrere der genannten Unterschiede wirken in die gleich Richtung und haben in Großbritannien die Entwicklung eines professionellen Immobilienmarktes mit unterschiedlichen Unternehmensformen und Anlagemöglichkeiten begünstigt. Dementsprechend ist die Bedeutung der privaten Anleger sowie die Bedeutung des Marktes für eigengenutzte Immobilien geringer als in der Bundesrepublik.

Wie die Darstellung zeigt, haben die Unterschiede des Steuerrechts eine zentrale Bedeutung. Die einzelnen steuerlichen Regelungen beruhen jeweils auf sehr starken Interessen, wobei das Vorhandensein der Regelungen, insbesondere das Vorhandensein von Steuervergünstigungen (Steuerbefreiung von realisierten Wertsteigerungen) sich im Laufe der Zeit eine eigene starke Interessenlobby schafft. Angleichungen in diese Richtung, sind auch im Zuge der EG-Integration wenn überhaupt nur allmählich zu erwarten. Die Politik hat demgegenüber einen größeren Gestaltungsspielraum, wenn es darum geht, das Entstehen professioneller Marktorganisationen, bessere Informationssysteme oder die Verbreitung besserer Ausbildungsleistungen zu unterstützen und zu fördern. Hier sind verschiedene Maßnahmen möglich, die für sich genommen die Marktprozesse nicht entscheidend verändern werden, die jedoch auch in der Bundesrepublik zu beobachtende Entwicklung in Richtung auf stärker professionelle Märkte beschleunigen können.

TEIL B
Case studies

Einzelhandel – Die Rolle
randstädtischer Einkaufszentren

Planungsstrategien

In Großbritannien besteht gegenüber großflächigen Einzelhandelszentren im Umland der Großstädte ähnlich wie in der Bundesrepublik eine restriktive Planungspraxis. Allerdings kann man in den letzten Jahren eine deutliche Tendenz zur Auflockerung dieser zurückhaltenden Genehmigungspolitik beobachten. Historisch lassen sich folgende Phasen der Entwicklung des Einzelhandelsmarktes unterscheiden:[9]

Die erste Dezentralisierungswelle betraf insbesondere den Lebensmitteleinzelhandel. Während der 70er Jahre eröffneten fast alle größeren Supermarktketten Filialen außerhalb der Innenstädte und suchten hierfür zumeist Standorte auf der 'grünen Wiese'. Diese erste Dezentralisierungswelle ist heute weitestgehend abgeschlossen. Zwar zeigen die suburbanen Supermärkte auch weiterhin ein deutliches Wachstum, aber die Diskussion um ihre Bedeutungen als Konkurrenz für den Lebensmitteleinzelhandel in den Stadtkernen ist inzwischen fast völlig verstummt. Heute wird der Verlust des Lebensmitteleinzelhandels für die Stadtzentren von vielen Planern sogar begrüßt, da sich somit viele Verkehrs- und Parkraumprobleme besser lösen lassen.

Die zweite Dezentralisierungswelle begann etwa 5 bis 10 Jahre nach der ersten und erreichte Mitte der 80er Jahre ihren ersten Höhepunkt. Sie betrifft vor allem den Bau von sogenannten 'retail wharehouses', in denen vor allem sperrige Güter wie Möbel und Teppiche angeboten werden. In den letzten Jahren entstanden zunehmend Heimwerkermärkte als neues Element des Einzelhandels. Auch diese sind kaum auf einen Standort im Stadtzentrum angewiesen oder dafür geeignet. Auch das Entstehen von 'retail warehouses' wird kaum als Bedrohung des innerstädtischen Einzelhandels angesehen, da hier die für den Verkauf sperriger Güter notwendigen Flächen noch zu tragbaren Preisen zur Verfügung stehen. Da in den Stadtzentren zudem kaum ausgedehnten Parkflächen bereitgestellt werden können, ist die suburbane Zone für viele dieser Einzelhandelsbetriebe die einzige Standortalternative.

Für die Vitalität des Stadtzentrums bedeutete keine der beiden ersten Dezentralisierungswellen eine ernsthafte Gefahr. Vielmehr konnten diese ihre starke Stellung als Schwerpunkte des Angebots für Waren des aperiodischen Bedarfs sogar noch ausbauen. Dies gilt insbesondere für den Bereich Textil und Bekleidung, der seit jeher die tragende Säule der Innenstadteinzelhandels darstellte. Ähnlich wie in der Bundesrepublik vollzog sich auch in den Zentren der britischen Städte eine ausgesprochene 'Textilisierung'. Insbesondere das Angebot an modischer Kleidung blieb lange Zeit fast ausschließlich auf den Cityeinzelhandel begrenzt. Deshalb löste 1984 die Ankündigung des größten britischen Bekleidungsspezialisten 'Marks and Spencer' nun auch 'out of town'-Filialen zu eröffnen, unter Planern und Einzelhändlern einen gehörigen Schock aus.

Dieser Zug des Unternehmens 'Marks and Spencer' markiert den Beginn der dritten Dezentralisierungsphase im britischen Einzelhandel. Anders als die beiden ersten Phasen, betrifft er vorwiegend jene Branchen, die traditionell in den Stadtzentren zu finden waren. Neben 'Marks and Spencer' engagieren sich inzwischen auch Unternehmen wie 'Habitat', 'Laura Ashley', 'Toys R Us' oder 'World of Leather' zunehmend an 'out of town'-Standorten. Diese Standorte treten somit in direkte Konkurrenz zu den Stadtzentren. Nirgends wird dies deutlicher als im Falle der 'regional centres', dieser großflächen Einkaufskomplexe, die seit Mitte der 80er Jahre am Rande mehrerer britischer Städte entstehen. Neben diesen, im Angebotsspektrum mit den Stadtzentren durchaus vergleichbaren 'regional centres', tritt in Großbritannie zunehmend auch die spezialisiertere Form des 'retail parks' auf. Hierbei handelt es sich zumeist um aus Gewerbegebieten entstandenen Standortkonzentrationen des großflächigen Einzelhandels.

Enterprise Zones als Gebiete besonderen Rechts

Als Folge der hohen Arbeitslosigkeit und des industriellen Zusammenbruchs zu Beginn der 80er Jahre wurden verschiedene Instrumente/Subven-tionsformen eingesetzt, um private Investitionen anzuregen. Die Enterprise Zones stellen den wohl radikalsten Versuch dar, private Investitionen in Problemgebieten anzuregen. Wichtigstes Element der Enterprise Zones ist das Fehlen von planerischen Restriktionen, d.h. Investoren haben das Recht, die Investitionen zu verwirklichen, die ihren privaten Zielen entsprechen. Hier werden nur in sehr weiten Grenzen Planungskontrollen wirksam (Beispiel: Kraftwerke oder andere gefährliche Anlagen..). Für den Investor war mit der Investition in einer Enterprise Zone eine zehnjährige Befreiung von den «rates» und eine Sofortabschreibung aller Anlageinvestitionen verbunden. Zur Überraschung vieler wurden in den Enterprise Zones zum Teil kaum industrielle Investitionen, dafür aber sehr große Einzelhandelszentren angeregt.[10]

Das Metrozentrum in Gateshead (nahe Newcastle)[11]

Das bekannteste Einkaufszentrum in einer Enterprise Zone entstand seit 1986 in Gateshead. Es ist in der Zwischenzeit zum größten Einkaufszentrum Europas gewachsen. Dabei kam es zu einer ständigen Nutzungsdifferenzierung. Das heutige Angebot geht weit über die reine Einzelhandelsfunktion hinaus. Das heute als 'shopping and leisure city' vermarktete Metro Centre bietet neben mehr als 300 Einzelhandelsgeschäften auch zahlreiche Freizeiteinrichtungen. Die Einzelhandelgeschäfte des Zentrums sind werktäglich bis 20 Uhr geöffnet, an Donnerstagen sogar bis 21 Uhr. Demgegenüber sind die Mehrzahl der Freizeiteinrichtungen, wie etwa der MetroLand-Vergnügungspark und der Kinokomplex bis spät in die Nacht geöffnet.

Lage und Verkehrsanbindung
Das Metro Centre liegt knapp 5 km südwestlich des Stadtzentrums von Newcastle, dem

traditionellen Zentrum der nordost-englischen Region 'Tyne & Wear'. Es ist sowohl über die Straße als auch über die Schiene aus der ganzen Region gut zu erreichen. Der Anschluß an das überörtliche Straßennetz erfolgt durch einen eigenen Anschluß an den A 69 Gateshead Western Bypass. Rund 1,4 Mio. Menschen können das Einkaufszentrum innerhalb einer Pkw-Fahrzeit von maximal 30 Minuten erreichen. Der sich derzeit im Bau befindliche Newcastle Western Bypass wird die Erreichbarkeit des Zentrums weiter verbessern.

Der Anschluß an das öffentliche Verkehrsnetz erfolgt im wesentlichen über Busse, mit denen etwa 3/4 der ÖPNV-Nutzer das Metro Centre erreichen. Neben normalen Linienbussen verkehren auch spezielle 'shuttle bus services', die das Metro Centre mit einer Fahrzeit zwischen 8 und 12 Minuten non-stop mit den Stadtzentren Newcastle und Gateshead verbinden. Als einziges 'neues' Einkaufszentrum in Großbritannien erhielt das Metro Centre 1987 einen eigenen, von der British Rail bedienten Eisenbahnhaltepunkt. Dieser ist zusammen mit dem Parkplatz für Reisebusse über einen überdachten Zugang direkt mit dem Zentrum verbunden.

Die Entwicklung des Metro Centres 1986-1991

Die ersten Vorüberlegungen zum Bau eines Einkaufszentrums, auf der vorher von einem Kraftwerk zur Ascheausbringung genutzten Fläche, gehen auf das Jahr 1980 zurück. Mitte 1981 wurde das 45 ha-Grundstück Teil einer Enterprise Zone, was seine Attraktivität für Einzelhandelsprojekte weiter steigerte. Insbesondere war mit der Ausnahme von Lebensmittelsupermärkten von über 250 m^2 Grundfläche für Einzelhandelsvorhaben keine weitere Genehmigung notwendig. Im Frühjahr 1984 stellte Cameron Hall erstmals seine Pläne für ein voll überdachtes regionales Einkaufszentrum vor. Diese stießen sowohl bei Einzelhändlern als auch bei öffentlichen Stellen auf großes Interesse.

Das Metro Centre wurde im April 1986 mit der Beendigung des ersten Bauabschnitts eröffnet. Es bestand zunächst lediglich aus einem Carrefour Supermakt sowie einigen kleineren Einzelhandelsgeschäften mit einer Gesamt-Nettogeschoßfläche von 7000m^2. Die zweite Bauphase wurde bereits im Oktober 1986 eröffnet und führte zu einer gewaltigen Vergrößerung der Einzelhandelsfläche auf nun zwei Stockwerken. Als zentrale Einzelhandelsgeschäfte eröffneten neben dem auf 12.500 m^2 vergrößerten Carrefour Supermarkt nun 'Marks and Spencer' (17.400 m^2 Bruttogeschoßfläche) und das 'House of Fraser' (12.900 m^2 Bruttogeschoßfläche). Die dritte Bauphase, welche sowohl den Kinokomplex als auch den Indoor-Freizeitpark umfaßte, wurde im Oktober 1987 eröffnet. Anfang 1989 wurde das Angebot an Freizeiteinrichtungen noch durch eine Bowlinganlage und ein Fitnesszentrum ergänzt. Somit besteht das Metro Centre heute im wesentlichen aus dem zweistöckigen Hauptkomplex mit mehreren überdachten 'shopping malls', der von ausgedehnten Parkplatzflächen und einer Ringstraße umgeben ist. Westlich des Hauptkomplexes wurden 1987 ein kleineres 'retail warehouse' und im Frühjahr 1990 ein Hotel-Büro-Komplex (150 Betten) fertiggestellt. Noch immer bestehen beachtliche Baulandreserven innerhalb des Geländes, insbesondere nördlich der Eisenbahnlinie. Anfang 1990 wurde eine Baugenehmigung für einen freistehenden Lebensmittelsupermarkt im nordwestlichen Teil des Geländes erteilt.

Nach einer Studie des Gateshead Metropolitan Borough Councils[12] umfaßte das Zentrum im Januar 1990 insgesamt 83.700 m^2 reine Einzelhandelsfläche. Hiervon entfielen 66.562 m^2 auf den Hauptkomplex und 17.138 m^2 auf das 'retail warehouse' (alle Angaben = Nettogeschoßfläche). Von dieser Fläche entfielen etwa 80% auf das

Angebot von langlebigen Konsumgütern, 7% auf Güter des kurzfrsitigen Bedarfs und 11% auf konsumorientierte Dienstleistungen. Die restlichen 2% der Fläche standen leer oder waren momentan nicht genutzt. Diverse Freizeit- und Vergnügungseinrichtungen ergänzen den Geschoßflächenbestand des Zentrums um weitere 15.000 m².

Anfang 1990 bestanden im Metro Centre insgesamt 318 Einzelhandelsgeschäfte. Hiervon waren etwa 55% Einzelbetriebe oder Filialen ausschließlich lokal agierender Einzelhändler. Dennoch entfielen auf diese kleineren Einzelhandelsunternehmen nur rund 10% der Gesamtgeschoßfläche des Zentrums.

Seit der Eröffnung im Jahre 1986 veränderte das Metro Centre seinen Charakter immer weiter in Richtung eines umfassenden Einkaufs- und Freizeitkomplexes. So wuchs nicht nur das den Einzelhandel ergänzende Angebot an Freizeit-, Erholungs- und Sporteinrichtungen, sondern man unternahm auch verstärkt den Versuch, den Einkauf selbst zum Mittelpunkt des Freizeitvergnügens zu machen. Neben zahlreichen allgemeinen architektonisch/gestalterischen Attraktivitäten diente hierzu vor allem die Austattung des Zentrums mit speziellen Bereichen thematisch einheitlicher Gestaltungsmerkmale. Beispiele für derartige 'Themenkomplexe' im Metro Centre sind etwa das Antique Village, das Forum Romanum oder das Mitte 1989 eröffnete Mediterranean Village. Der in das Zentrum integrierte MetroLand-Vergnügungspark ist der erste seiner Art in Europa. Dasselbe gilt für den 2500 Personen fassenden Kinokomplex und den 40 Restaurants und Cafes umfassenden zentralen 'food court'.

Das wachsende Freizeitangebot innerhalb des Metro Centres hat in den letzten Jahren sicher stark zu dessen wachsender Attraktivität beigetragen. Dies gilt insbesondere in bezug auf Familien mit Kindern. Zudem erhöhte sich durch das Angebot an Freizeiteinrichtungen auch der Umsatz des Einzelhandels und der gastronomischen Einrichtungen. Etwa drei Viertel der Besucher der Freizeiteinrichtungen gaben während ihres Besuchs im Metro Centre auch Geld in den Einzelhandelsgeschäften aus.

Das Metro Centre als Arbeitsplatz

Durch sein rasches Wachstum wurde das Metro Centre ein bedeutender regionaler Beschäftigungsschwerpunkt. Anfang 1990 waren im Metro Centre insgesamt 5370 Personen beschäftigt. Die Mehrzahl der Beschäftigten waren Frauen (72%). Rund die Hälfte der Arbeitsplätze boten lediglich Teilzeittätigkeiten (51%). Dies gilt insbesonere für die Arbeitsplätze im Einzelhandel und Gastgewerbe (etwa 92% aller Arbeitsplätze). Demgegenüber dominieren Männer (76%) auf den meist höherqualifizierten Positionen im Management und im Sicherheitsdienst (etwa 3% aller Arbeitsplätze). Etwa die Hälfte aller Arbeitskräfte wohnen in Gateshead selbst und etwa ein weiteres Drittel kommt aus der übrigen Region 'Tyne & Wear'. Dies unterstreicht die nicht zu unterschätzende Bedeutung des Metro Centres für den lokalen Arbeitsmarkt.

Die Verkehrsituation

Mit einem Parkplatzangebot von rund 10.000 Stellplätzen ist das Metro Centre trotz seiner vergleichsweise guten Anbindung an das ÖV-Netz hauptsächlich auf Einkaufsfahrten mit dem Privat-Pkw ausgerichtet. Eine im Juni 1990 durchgeführte Untersuchung ergab, daß etwa 79% der Kunden mit dem eigenen Pkw anreisen, während nur 16% den Bus und gerade 4% die Bahn benutzen. Erwartungsgemäß kommt nur ein sehr kleiner Teil der Kunden zu Fuß oder mit dem Fahrrad (1%) zum Zentrum.

Günstiger stellte sich der 'modal split' bei den im Zentrum Beschäftigten dar. Von dieser Personengruppe benutzen etwa 41% den privaten Pkw aber rund 57% öffentliche Verkehrsmittel für den Weg zur Arbeit. Von den ÖV-Fahrten entfällt das Gros der Fahrten auf Busse (89%), während die Bahn als Transportmittel auch hier eine deutlich geringere Rolle spielt (11%).

Ein Einkaufszentrum von der Größe des Metro Centres zieht beachtliche Verkehrsmengen an. Seit der Eröffnung der zweiten Bauphase 1986 unterliegt der Verkehrsfluß zum Metro Centre einem ständigen Monitoring. Das Verkehrsaufkommen zeigt bei saisonalen Schwankungen seither einen deutlichen Aufwärtstrend. Spitzenwerte werden erwartungsgemäß v.a. in der Woche vor Weihnachten erreicht. In der Vorweihnachtswoche 1989 wurden an sechs Tagen 200.500 Kraftfahrzeuge gezählt. Dies waren 11% mehr als noch im Jahr davor. In solchen Spitzenphasen treten insbesondere an den Zufahrten zum Zentrum bereits erste Überlastungserscheinungen auf. Um diesen Spitzebenlastungen Herr zu werden, wird seit 1989 in der Vorweihnachtswoche ein spezielles Verkehrsmanagement angewandt.

Die starke Abhängigkeit von der Attraktivität für 'Autokunden' ist einer der immer wieder vorgebrachten Hauptkritikpunkte an der Verbreitung freistehender Einkaufszentren. Freistehende suburbane Einkaufszentren bleiben ab einer bestimmten Größe nicht ohne einen – zumeist negativen – Einfluß auf das städtische und regionale Verkehrsgeschehen. Dies wiegt umso schwerer, da gleichzeitig die Gefahr einer Unterauslastung der öffentlichen Verkehrsinfrastruktur in den etablierten Stadtzentren besteht. Gerade in diesen Bereich flossen aber jahrzehntelang beachtliche Summen öffentlicher Investitionen.

Die Auswirkungen des Einkaufszentrums auf Newcastle und die umliegenden Städte

Durch seinen Angebotsschwerpunkt bei Waren des mittel- und langfristigen Bedarfs tritt das Metro Centre in direkte Konkurrenz zu den bereits etablierten Zentren der Region, insbesondere zu den naheliegenden Stadtzentren von Newcastle und Gateshead. Tatsächlich ist das Metro Centre heute ein Einkaufszentrum mit gesamtregionaler Bedeutung und weist einen dementsprechend weiten Einzugsbereich auf. Zwar kommen etwa 55% der Einkäufer aus einer Zone mit bis zu 15 Minuten Fahrzeit vom Zentrum, aber fast 1/5 der Kunden nimmt auch Anfahrten von mehr als 30 Minuten in Kauf. Als Touristenattraktion bekommt das Zentrum selbst Besucher aus recht entfernten Regionen, die zum Teil dazu bereit sind, bis zu drei oder mehr Stunden Anfahrt zu akzeptieren. Mit zunehmendem Alter des Metro Centres und einer weiteren Verbreitung von gleichartigen Einkaufszentren wird deren Zahl aber in Zukunft vermutlich merklich zurückgehen.

Das Besucherprofil des Metro Centre unterscheidet sich von dem anderer Zentren der Region durch einen überdurchschnittlichen Anteil junger, aber relativ wohlhabender Familien sowie Vorortsbewohnern mittleren Alters. Dies gilt insbesondere gegenüber dem Stadtzentrum von Newcastle, das offensichtlich eine starke Attraktivität für die Bewohner innerstädtischer und besonders exklusiver Wohngebiete hat.

Im Juni 1989 lag er durchschnittliche Betrag der pro Einkäufer und Tag im Metro Centre ausgegeben wurde bei rund £46. Dies war eine Steigerung von etwa 12% gegenüber dem Vorjahr. Damit gaben die Einkäufer im Durchschnitt pro Besuch im Metro Centre mehr Geld aus, als in anderen Zentren innerhalb der Region.

Insbesondere beim Angebot langfristiger Konsumgüter hat das Metro Centre einen starken Einfluß auf die regionale Einzelhandelsstruktur. Schätzungen gehen davon aus, daß der Jahresumsatz des Zentrums bei langfristigen Konsumgütern mit mehr als £300 Mio. (1989) etwa ein Drittel des in diesem Sektor erbrachten Gesamtumsatzes innerhalb des 30-Minuten-Einzugsbereichs darstellt.

Innerhalb der Gemeinde Gateshead erlebte das Stadtzentrum selbst die stärksten Einbußen im Einzelhandel. Innerhalb der letzten Jahre reduzierte sich die Gesamt-verkaufsfläche merklich durch die Schließung zweier großer Einzelhandelsgeschäfte, von denen allerdings keine in einem direkten Zusammenhang mit der Eröffnung des Metro Centres stand. Starke Auswirkungen wurden ebenso auf den Einzelhandel in der Newcastler Innenstadt erwartet. Diese Befürchtungen haben sich bislang nicht im erwarteten Maße erfüllt. Insbesondere blieben stärkere Umsatzeinbußen in Newcastle und den anderen etablierten Zentren bis heute auch deshalb weitestgehend aus, weil das Metro Centre zu einer Zeit eröffnet wurde, in der sich die Einzelhandelsumsätze generell gut entwickelten. Eine Schätzung des Newcastle City Council geht davon aus, daß während der ersten beiden Jahre, in denen das Metro Centre in Betrieb war, der Umsatz in Newcastle dennoch um etwa 3 bis 8% abgenommen hatte. Seit der Eröffnung des Metro Centres stieg die Gesamtverkaufsfläche in der Innenstadt von Newcastle jedoch weiter an, ohne daß merkliche Leerstandprobleme auftraten.

Insgesamt hat das Entstehen des Metro Centre die regionale Einzelhandelshierarchie merklich verändert. Vor allem reduzierte sich die traditionell ausgeprägte Dominanz Newcastles innnerhalb der Region 'Tyne & Wear'. Dabei ist das Einzelhandels- und Dienstleistungsangebot des Zentrums zu heterogen, um es selbst eindeutig einer Hierarchiestufe zuordnen zu können. So hat das Metro Centre zwar einen verhältnis-mäßig großen Einzugsbereich, durch die Nähe Newcastles fehlt ihm aber das für ein 'echtes' regionales Einkaufszentrum notwendige Marktdurchdringung. Dies gilt insbesondere für den urbanen Kernbereich der Region.

Der Meadowhall-Komplex in Sheffield[13]

Ähnlich wie in Gateshead entstand auch in Sheffield ein vollüberdachter Einkaufs-Freizeit-Komplex von gigantischen Ausmaßen. Die Ende 1990 eröffnete Meadowhall bietet auf zwei Etagen etwa 120.000 m² Fläche für über 230 Einzelhandelseinheiten und einen von 7 Uhr morgens bis 2 Uhr nachts geöffneten Freizeitbereich. Das Zentrum befindet sich im Lower Don Valley in etwa 3 km Entfernung von Sheffields Stadt-zentrum. Früher befand sich auf dem Gelände ein großes Eisen- und Stahlwerk, das Anfang der 80er Jahre endgültig geschlossen wurde. Obwohl der Sheffield City Council auf dem Gelände ursprünglich industrielle Nutzungen unterbringen wollte, führten folgende Punkte zur Genehmigung und aktiven Unterstützung des Einzelhandelsprojekts:
- Sowohl im Stadtgebiet als auch in der Region bestand Mitte der 80er Jahre eine Unterausstattung mit Einzelhandelsflächen.
- Von dem Einkaufszentrum versprach man sich als 'Attraktivitätspol' einen positiven Effekt auf die weitere Entwicklung der umliegenden Flächen.
- Das Grundstück bot durch seinen günstigen Verkehrsanschluß und seine zentrale Lage innerhalb der Region sehr gute Voraussetzungen zur Entwicklung eines Einkaufs-zentrums.

Tatsächlich ist die Lage sowohl im lokalen als auch im regionalen Kontext nahezu ideal. So kann das Zentrum innerhalb einer halben Stunde Fahrzeit von rund 2,2 Millionen Menschen erreicht werden. Es liegt unmittelbar an einer Ab- und Auffahrt des M1-Motorways und besitzt einen eigenen Eisenbahnhaltepunkt sowie einen eigenen Busbahnhof. Der Bau eines neuen Supertram-Systems, welches das Zentrum direkt mit der Innenstadt von Shffield verbinden soll, befindet sich in Planung. Wie das Metro Centre in Gateshead, so zielt auch die Meadowhall vorwiegend auf 'Autokunden', für die rund 12.000 kostenlose Parkplätze zur Verfügung stehen. Zusätzlich befinden sich auf dem Gelände 300 Stellplätze für Reisebusse.

Es ist derzeit noch verfrüht, Aussagen über die Auswirkungen des Meadowhall-Komplexes auf andere Einzelhandelszentren der Region zu machen. Eine im Vorfeld angefertigte Studie geht davon aus, daß das Einkaufszentrum Mitte der 90er Jahre den Umsatz für langfristige Konsumgüter in Sheffields Stadtzentrum um etwa 12–13% senken könnte. Bislang wurden in Sheffield etwa 50% des Umsatzes im non-food-Bereich im Stadtzentrum erzielt.

In jüngerer Zeit ist es zu Unstimmigkeiten zwischen dem Projektträger und den zuständigen Planungsbehörden gekommen, da insbesondere die Qualität des Freizeitangebotes innerhalb des Zentrums nicht die Erwartungen der Behörden erfüllte. Eine vom Developer beabsichtigte Erweiterung der Einzelhandelsfläche wurde daraufhin nicht genehmigt.

Fazit/Folgerungen für die Bundesrepublik

In der Bundesrepublik besteht eine sehr restriktive Praxis bei der Ausweisung großflächigen Einzelhandels. De facto werden Großprojekte einer Bedürfnisprüfung unterzogen. Dabei wird geprüft, ob bestehende Zentren, insbesondere Innenstadtzentren, in ihrer Funktionsfähigkeit beeinträchtigt werden. Diese Praxis unterdrückt eine latente Nachfrage nach großen komplexen Zentren, in denen nicht nur Einkaufsmöglichkeiten sondern auch Freizeitmöglichkeiten bestehen. Dort, wo in Europa planerische Restriktionen nicht bestehen, haben sich i.d.R. solche großen Zentren entwickelt (Austriazentrum in Wien, Amoreiras in Lissabon . . .). In der Bundesrepublik, nochmehr in den Niederlanden, bestehen allerdings große spezialisierte Freizeiteinrichtungen (Bad Soltau, Phantasialand . . .). Dadurch, daß große Kombinationszentren nicht entstehen können, wie sie der Nachfrage von zahlreichen Haushalten entsprechen, wird im Ergebnis den Wettbewerb reduziert. Nachfrage wird in die Stadtzentren umgelenkt, mit allen Konsequenzen (hohe Verkehrsbelastungen in den Zentren, scharfer Auslesewettbewerb mit der Folge der Vereinheitlichung der Angebote und der Verdrängung von weniger leistungsstarken Spezialangeboten). Gleichzeitig wird Einkaufen als Freizeitbeschäftigung in seiner Entfaltungsmöglichkeit behindert. Wie man aus Beobachtung des Konsumentenverhaltens in den großen kombinierten Zentren wie dem Metro Centre Gateshead weiß, verbringen die Besucher dort oft mehrere Stunden. Es werden Waren und Angebote verglichen. Neben dem Einkaufen werden auch Freizeitangebote in Anspruch genommen.

Die für die Stadtplanung spannende Frage lautet, ob die Rechtfertigungen, die für eine Beschränkung des Wettbewerbs und die Eingrenzung der Wahlfreiheit von Kunden gefunden werden, tatsächlich ausreichen. Dabei liegt auf der Hand, daß Wirkungsanalysen solcher großen Zentren nur mit erheblichem Aufwand durchgeführt werden können und darüberhinaus immer mit gewissen Unsicherheiten behaftet sind. Insbesondere ist schwierig festzulegen, wo die Wirkungsketten abzubrechen sind. Die in der Bundesrepublik üblicherweise vorgetragenen Argumente zugunsten einer restrikten Planungspraxis konzentrieren sich fast ausschließlich auf die unmittelbar beobachtbaren Wirkungen. Dabei treten, wie man am Beispiel von Gateshead feststellen kann, sehr komplexe langfristige Wirkungen auf.

Beispiele:
- Unter den Versagungsgründen, die üblicherweise in der Bundesrepublik gegenüber großen, nicht integrierten Umlandzentren angeführt werden, wird die Verringerung der Einzelhandelsumsätze in bestehenden Zentren als wichtiges Argument angesehen. Nicht berücksichtigt wird, welche alternativen Nutzungen bei einem Rückgang des Nachfragewettbewerbs in den Stadtzentren oder anderen innerstädtischen Zentren Platz finden würden. So würden mit hoher Wahrscheinlichkeit gegenwärtig in disperse Standorte abgedrängte schwächere Nutzungen in zentralere Standorte zurückkehren. In den entlasteten Stadtkernen würden sich andere Nutzungmischungen herausbilden. Eine Gesamtbewertung längerfristiger Wirkungsketten ist auf einer rein hypothetischen Ebene überhaupt nicht möglich. Empirische Wirkungsketten können angesichts fehlender Beispiele in der Bundesrepublik nicht untersucht werden.
- Auch der Hinweis auf den zusätzlichen Einkaufsverkehr konzentriert sich nur auf einen ersten Wirkungsschritt. Dieser Verkehr tritt allerdings in erheblichem Umfang zu Zeiten auf, in denen der sonstige Wirtschaftsverkehr relativ niedrig ist. Die Kombination aus Freizeit- und Einkaufsverkehr wirkt insgesamt wieder verkehrsvermindernd. Die Auswirkungen auf das Stadtzentrum von Newcastle scheinen nach bisherigen Informationen nicht sehr gravierend zu sein, d.h. ein größerer Teil der Kaufkraft wird natürlich außerhalb des Zentrums ausgegeben. Damit entfallen Angebotskombinationen wie sie in Zentren unter restriktiven Bedingungen zu beobachten sind. Dies bedeutet aber nicht, daß die entsprechenden Geschäfte leer werden. Sie werden von anderen Funktionen besetzt. Dies bedeutet weniger Umsätze. Dadurch können sich Geschäfte in den Zentren halten, die bei hohem Wettbewerbsdruck in disperse Standorte abgedrängt würden. Die Bindung des Angebots im Umland geht zu Lasten von dispersen Standorten innerhalb der Kernstadt einschließlich ihrer Randzonen. In der Kernstadt werden dadurch auch stärker freizeitorientierte Dienstleistungsfunktionen möglich. Im Ergebnis ist offen, ob die entstehende Nutzungskombination nicht zu einer attraktiveren Innenstadt führt, als die durch hohen Wettbewerbsdruck vereinheitlichten Stadtzentren. Selbst bei den Verkehrsbelastungen sind die Gesamtwirkungen wahrscheinlich differenzierter als die unmittelbare Wirkung der Verkehrsbelastung durch den Besuch der großen Zentren selbst signalisieren. So werden z.B. andere Streunutzungen stärker in den Kernstädten konzentriert, was dort wiederum den Verkehr bündelt. Dem steht entgegen, daß aus den Zentren der Umlandgemeinden Verkehr auf die großen nicht integrierten Zentren umgelenkt werden. Allein diese wenigen Hinweise machen deutlich, wie komplex die Wirkungen sind. In Großbritannien hat das Department of Environment eine Evaluation der Auswirkungen der großen nicht integrierten Zentren gestartet. Zur Zeit der Befragungen zu dieser Studie lagen die Ergebnisse noch nicht vor. Es empfiehlt sich, sie nach ihrer Veröffentlichung auszuwerten.

2
Rolle der Welsh Development Agency als Immobilienentwickler

Aufgaben der Development Agencies

Die Scottish Development Agency genau so wie die Welsh Development Agency sind Ergebnis der besonderen regionalen Autonomie, die Schottland und Wales in Großbritannien politisch genießen. Die Welsh Development Agency wurde wie die Scottish Development Agency (1976) gegründet. Sie übernahm bei ihrer Gründung das Industrievermögen der Regionalregierung in Wales und wurde beauftragt, daß «Derelict-Land-Reclamation-Programme» weiterzuführen. Wichtiger waren jedoch die Aufgaben, künftig durch eigene Investitionen und Beratung und durch Akquisition von Ansiedlungen die wirtschaftliche Entwicklung von Wales zu fördern. Dabei erhielt die Welsh Development Agency auch die Möglichkeit, als Vermieter von gewerblichen Gebäuden und Büros aufzutreten. Sie konnte sich in Extremfällen auch mit Eigenkapital an der Ansiedlung und Neugründung von Unternehmen beteiligen. Sie ist eine Organisation, die es in dieser Kombination in der Bundesrepublik nicht gibt, weil sie Wirtschaftsförderungsaufgaben mit den Aufgaben einer Landesentwicklungsgesellschaft verbindet. Die Scottish Development Agency wird gegenwärtig aufgrund von Beschlüssen der Regierung in London umstrukturiert. Die bisher als staatliche Leistungen angebotenen Dienste sollten in Zukunft stärker «vermarktet» werden. Die Scottish Development Agency (künftig Scottish Enterprise) zieht sich insbesondere als Vermieter weitgehend vom Markt zurück.

Die wirtschaftliche Entwicklung als Hintergrund

Die 70er und frühen 80er Jahre waren in Wales geprägt durch einen rapiden Niedergang der alten Industrien. Dies hatte erhebliche Freisetzungen von Grundstücken zur Flge. Angesichts des wirtschaftlichen Niedergangs kam es jedoch zu keinen nennenswerten privaten oder kommunalen Grundstücksentwicklungen. Hierbei ist zu berücksichtigen, daß die Kommunen in Großbritannien bei der Entwicklung von Grundstücken auch für Industriezwecke eine sehr viel schwächere Rolle übernehmen als die privaten Entwicklungsgesellschaften. Viel häufiger als in der Bundesrepublik werden die entsprechenden Grundstücke von privaten Investoren aufbereitet und vermarktet. Dabei werden sie vielfach bebaut vermietet. Es gibt einen großen vermieteten Bestand an gewerblichen Objekten. Die Entwicklung in Wales war in der zweiten Hälfte der 80er

Jahre durch ein kräftiges Wachstum der Gesamtbeschäftigung (1987 1,032 Mio.; 1990 1,236 Mio.) gekennzeichnet. Dabei kam es zu einer erheblichen Differenzierung der ehemals sehr einseitigen Wirtschaftsstruktur. Neben einem überdurchschnittlichen Wachstum von High-Tech-Industrien expandierte auch der Dienstleistungssektor (Finanzdienstleistungen 1984: 2.000 Beschäftigte; 1990: rd. 6.500). Die Welsh Development Agency schreibt sich, ohne daß dies im Einzelfall nachweisbar wäre, einen erheblichen Anteil an diesen Erfolgen zu. Dabei zeigen zahlreiche Beispiele, daß in der Kombination verschiedener Instrumente, die in der Bundesrepublik nicht in einer Hand zusammengefaßt sind, erhebliche Wirkungsmöglichkeiten liegen. Die Welsh Development Agency schreibt über sich selbst: «The WDA has a very substantial holding of industrial land and advance factory premises throughout Wales, including some which have been specifically developed for the new technologies. Alternatively, the WDA can design and build premises to meet the needs of individual businesses or can offer fully serviced sites at strategic locations. Sites and premises are available for lease or sale. However, if a private sector route is preferred, Welsh Development International can establish contacts with appropriate landowners, developers and construction companies . . . Manufacturing establishments and certain qualifying service industries who set up or expand in the Assisted Areas (of which Wales contains a substantial proportion) may obtain Government grant aid in the form of Regional Selective Assistance. Grant is based on the capital expenditure costs of the projects and the number of jobs involved. Welsh Development International staff work closely with the relevant government department in assessing grant applications and in speeding up their approval and disbursement. Welsh Development International and WINtech can also advise on other potential sources of grant aid associated, for example, with research and development activities to be undertaken by the joint venture.»[14] Die Kombination aus Grundstücksentwicklung und Wirtschaftsförderung, die auch die Kompetenz beinhaltet, über Gewährung von Investitionszulagen mitzuentscheiden, erlaubt eine Konzentration von Fördermaßnahmen auf besondere risikoträchtige Standorte. Pionierinvestoren, die in der Frühphase einer regionalen Entwicklung in noch unsicherem Umfeld investieren, können besonders hohe Förderanreize erhalten. Aus zahlreichen Äußerungen wird deutlich, daß die Unternehmen einen solchen «One Stop Shop» für sehr hilfreich halten.

Die Welsh Development Agency als Grundstücksentwickler

Angesichts der Angebotslücken in Wales übernahm die Welsh Development Agency schon sehr früh die Funktion eines Grundstücks- und Immobilienanbieters. Sie folgte in ihren Standortentscheidungen in der frühen Phase sehr stark dem jeweiligen Arbeitsmarkt, d.h. Standorte wurden dort entwickelt, wo Arbeitslosigkeit besonders hoch war. Dabei wurden Grundstücke in der Regel unter den Gestehungskosten verkauft. Die Differenz wird aus dem Landeshaushalt finanziert. Die Welsh Development Agency bietet aber auch gleichzeitig Gebäude zur Vermietung an und wurde ähnlich wie die Scottish Development Agency innerhalb weniger Jahre zum größten Anbieter neuer Gewerbeimmobilien und zum wichtigsten Vermieter.

Phasen der Grundstücksentwicklung

In der historischen Entwicklung kann man mehrere Phasen unterscheiden. Die, wie dargestellt, erste Phase ar stark entwicklungsorientiert und bezog sich sehr stark auf die Situation am Arbeitsmarkt. Im Laufe der 80er Jahre traten folgende Veränderungen ein:
- Die Thatcher-Regierung stand der Welsh Development Agency als einem staatlichen Monopolanbieter kritisch gegenüber. Von daher wurde Druck ausgeübt, den Anteil der vermieteten Immobilien zu reduzieren. Genau so wie es als weniger erwünscht galt, sich stark mit Eigenkapital an Neuansiedlungen zu beteiligen.
- Im Zuge des Aufschwungs nach 1983 erhöhte sich auch die Nachfrage nach Gewerbeimmobilien in Wales. Für private Investoren wurde ein Markt sichtbar.
- Die Erfahrungen mit der eigenen Vermarktung legten es nahe, stärker als in den Anfängen die Standortbedürfnisse der Investoren und Nachfrager zu berücksichtigen und nicht in erster Linie exogen politisch bestimmte Standortziele bei den Eigeninvestitionen zu berücksichtigen.

Als Ergebnis dieser drei Faktoren hat sich die Standortwahl der Entwicklungen verändert. Sie folgt mehr den Standortwünschen von Investoren. Daneben wurden seit Mitte der 80er Jahre stärker Joint-Venture-Unternehmen gestartet. In der Folge wurde eine eigene Grundstücksentwicklungsgesellschaft «Welsh Property Venture» gegründet (1989). Sie investiert fast regelmäßig zusammen mit den Kommunen und privaten Investoren. Die gemeinsamen Entwicklungen umfassen die gesamte Bandbreite der Immobilienwirtschaft (Bürobauten, flexible Bauformen, die sich für Büros und hochwertige Produktion eignen, Industrie- und Lagergebäude). Mehrere Gewerbeparks wurden errichtet. Insgesamt vermietet die Welsh Development Agency rund 240.000 m² an Gewerbe- und Büroflächen. Für 1991/92 ist ein Investitionsprogramm für 150.000m² geplant. Davon sollen rund 60.000 m² in Joint-Venture-Projekten entwickelt werden. Nach wie vor ist es Ziel der Beteiligung der Welsh Development Agency, durch Bereitstellung von zinsgünstigen Finanzierungsmitteln das Entwicklungsrisiko zu senken. Darüber hinaus soll natürlich für die Welsh Development Agency ein Ertrag erwirtschaftet werden. Rund die Hälfte ihrer Einnahmen stammen gegenwärtig aus eigenen Erlösen. Durch ein großzügiges Angebot sollen schließlich die Preise ganz generell niedrig gehalten werden. Insbesondere sollen «spekulative Situationen» vermieden werden.

Überdurchschnittliche Erfolge?

Eine Bewertung einer Grundstücksentwicklung ist sehr schwierig. Die Welsh Development Agency weist darauf hin, daß es gelungen ist in Wales insgesamt 20% der Ansiedlungen von internationalen Firmen (u.a. Bosch) zu erreichen. Das ist weit mehr als dem Bevölkerungsanteil entspricht. Diese Ansiedlungen sind natürlich nicht nur Ergebnis der Grundstückspolitik, sondern auch Ergebnis der anderen Anwerbemaßnahmen und des geringen Lohnniveaus in Wales. Die Welsh Development Agency vertritt das Konzept eines «One Stop Shop», d.h. ansiedlungswilligen Unternehmen bietet sie sich als Partner an, der für alle Fragen zuständig ist. Dabei werden Arbeitsmarkt- und Standortanalysen erstellt. Die Welsh Development Agency sucht den für das ansiedlungswillige Unternehmen günstigsten Standort. Sie ist bereit, den Standort zu entwickeln, zu erschließen und dem Unternehmen zur Verfügung zu stellen. Dabei übernimmt sie alle Planungs- und Infrastrukturleistungen. Gleichzeitig wird außerhalb Großbritanniens ein ständiger Informationsservice aufrecht erhalten. Gestützt auf diese Aktivitäten soll Wales zu einem attraktiven Ansiedlungsstandort entwickelt werden.

Stadterneuerung als spezielle Aufgabe

1986 wurde eine eigene Stadterneuerungsagentur ins Leben gerufen. Sie unterstützt die Gemeinden bei ihren Erneuerungsaufgaben. Dabei operiert sie anders als die Kommunen, weil sie entsprechend ihres Unternehmensauftrags die gewerbliche Entwicklung zu fördern hat. D.h., Stadterneuerung wird mit dem Ziel vorgenommen, Neugründungen und Ansiedlungen von Unternehmen zu begünstigen. Im Selbstverständnis der Gesellschaft muß sie nach «Market Opportunities» suchen, um eine «Enterprise Culture» zu fördern. Das bedeutet, bei der Vermarktung besthen kaum Unterschiede gegenüber Neuentwicklungen. Die Stadterneuerungsbemühungen unterscheiden sich lediglich in den Techniken und in den Gebietstypen in denen die neuen Grundstücke erschlossen und entwickelt werden.

Folgerungen für die Bundesrepublik

In der Bundesrepublik sind kombinierte Stadtentwicklungs- und Wirtschaftsförderungsgesellschaften aufgrund des Ressortprinzips der Landesministerien nie entstanden. Tatsächlich zeigen nicht nur britische Beispiele, daß in der Kombination beider Maßnahmen der eigentliche Erfolg liegt. In einem Gespräch im DOE wurde im Herbst 1991 angedeutet, daß die Regierung darüber nachdenkt, in mehreren Regionen ergänzend zu den zeitlich befristeten auf bestimmte Grundstücke konzentrierten Urban Development Corporations auf Dauer angelegte Entwicklungsagenturen zu gründen. Dabei scheint auch daran gedacht zu sein, für England insgesamt eine Entwicklungsagentur ins Leben zu rufen. Die Synergieeffekte, die möglich sind, in dem man allgemeine Wirtschaftsförderung mit Grundstücksentwicklung verknüpft führen grundsätzlich zu einer sehr viel günstigeren Ansiedlungssituation bei den Investoren. Schwer abzuschätzen ist die optimale Größe einer solchen Organisation. Gegenüber der Scottish Development Agency mit 600 Beschäftigten wurde immer wieder kritisiert, daß sie sich zu einer schwerfälligen Bürokratie entwickelt habe. Dies hat dort dazu geführt, daß eine eigene Akquisitions- und Wirtschaftsförderungsorganisation gegründet wurde (Lisc-Locate in Scotland). Allerdings übernimmt Lisc im wesentlichen nur die Akquisition in Übersee, d.h. nicht alle Wirtschaftsförderungsmaßnahmen werden in einer speziellen Organisation separiert. Die um zwei Drittel kleinere Welsh Development Agency scheint in ihrer internen Größe eine bessere Dimension zu haben. So kann die Welsh Development Agency trotz ihrer geringeren Größe auch im Ausland mehrere Akquisitionsbüros unterhalten (Tokio, Korea, Stuttgart). Was sich in jedem Fall bewährt hat, ist die Verzahnung von Wirtschaftsförderung und Grundstücksentwicklung. In Gebieten mit besonders hohem Entwicklungsrisiko für Grundstücke, wo sich Märkte zunächst gar nicht herausbilden, ist sie auch mit Risikokapital engagiert.

Folgerungen und Bewertungen

In der Bundesrepublik ist die Entwicklung anders verlaufen. Aufgrund der Ressort-spezialisierungen gibt es in den Bundesländern von der Grundstücksentwicklung abgetrennte Wirtschaftsförderungsorganisationen. Dies dürfte die Effizienz dieser Bemühungen generell reduziert haben. Darüber hinaus haben die Kommunen eine sehr starke Stellung bei der Entwicklung von industriellen Grundstücken. Ein hochentwickelter Markt für Gewerbegrundstücke außerhalb des Bürobereichs besteht nur in Ansätzen. Auch diese Entwicklung ist problematisch. Die Subventionierung von Ansiedlungen über günstige Grundstückspreise sollte nur unter zwei Zielen aufrecht erhalten werden:

– Wiedernutzung von Recyclinggrundstücken, deren Wiedernutzung sich privatwirtschaftlich nicht rentiert.

– Ansiedlungsförderung in Regionen mit hoher Arbeitslosigkeit, wo private Investoren sich nur selten engagieren.

Alle anderen Förderungen sind nicht zu rechtfertigen, insbesondere nicht im Hinblick auf die verschärften Umweltrestriktionen. Jede Flächenbereitstellung zu subventionierten Preisen fördert eine Tendenz zur Flächenhortung.

Da die Entwicklung in der Bundesrepublik nicht mehr zurückzudrehen ist, sollten sich die Bemühungen darauf konzentrieren, die Subventionen, die bei der Grundstücksvergabe implizit gewährt werden, zu reduzieren. Zu prüfen wäre allerdings, ob in Ostdeutschland, vor allem in den Risikogebieten, in denen die Arbeitslosigkeit noch über lange Jahre bestehen wird, nicht auf Erfahrungen Großbritanniens zurückgegriffen und Grundstücksentwicklung und Wirtschaftsförderung zusammengeführt wird.

3
Die Entwicklung von Canary Wharf

Vorgeschichte

Die Vorgeschichte ist vielfältig analysiert und dargestellt. Deshalb sind hier nur einige wesentliche Elemente in Erinneung zu rufen. Durch den Niedergang der DockFunktionen und der industriellen Produktion in den Docklands entstand in den 70er Jahren unmittelbar angrenzend an die City ein fast zwanzig Kilometer langes Band von Stadtbrache bzw. ein Gebiet mit extensiven Nutzungen (Lagerhaltung, einfache Fertigungen). Dementsprechend stieg die Arbeitslosigkeit in den Boroughs Ost-Londons lange Zeit auf über 20% an. Gegen diesen industriellen Niedergang versuchten die Regierungen der 70er Jahre zusammen mit den Boroughs mehrere Revitalisierungskonzepte. Alle diese Konzeptionen gingen von der Prämisse aus, daß in den Docklands eine Gewerbezone mit überwiegend nicht störender Fertigung entstehen sollte. Die Büroinvestitionen wurden jeweils nur als untergeordnete Nutzungen angesehen. Diese Konzeptionen sind einmal daraus zu erklären, daß versucht wurde für die Arbeitslosen und ihre Qualifikation «dazu passende» Arbeitsplätze zu schaffen, d.h. die Qualifikation und die Ausrichtung der Arbeitsplätze sollte sich an der Qualifikation und der bisherigen Berufserfahrung der Bevölkerung in Ost-London orientieren. Darüberhinaus wurde die Transformation Londons zu einem Dienstleistungszentrum nicht in dieser Radikalität gesehen.

In der öffentlichen Diskussion wird vielfach übersehen, daß es in den Docklands auch zu industriellen Entwicklungen kam. Dabei handelt es sich typischerweise um Betriebe, die direkt für den großstädtischen Markt der Region Londons produzieren. Am bekanntesten sind die Auslagerungen von Druckereien in die Docklands. Allerdings gilt auch hier, daß die neuen Industriebeschäftigten andere Qualifikationen benötigen als die ehemaligen Dockland-Arbeiter.

Neben den Versuchen einer Reindustrialisierung sollten große Wohngebiete mit sozialem Wohnungsbau entstehen. Dieses Revitalisierungskonzept hat sich nicht realisieren lassen. Einmal erwies es sich bis heute als fast unmöglich in innerstädtischen Standorten im größeren Stil neue Industrien anzusiedeln. Die Transportbedingungen sind dafür zu ungünstig und die Produktionskosten einschließlich der Lohnkosten zu hoch. Die Nähe zum Fluß bringt für moderne Industrien kaum einen Vorteil. Die Transportfunktionen der Themse läßt sich in ihrer alten Bedeutung nicht wieder herstellen. Unabhängig von diesen inhaltlichen Problemen stellt es sich heraus, daß eine Abstimmung zwischen drei Kommunen erhebliche Schwierigkeiten zur Folge hatte - jedenfalls wurde dies von der konservativen Regierung seit Ende der 70er Jahre immer wieder betont.

Als Folge der komplexen und wenig erfolgreichen Bemühungen wurde 1982 ein völlig neuer Ansatz versucht. Schon 1979 hatte der damalige Secretary of State for the Environment, Michael Heseltine, die Gründung von Urban Development Corporations für das Gebiet der Docklands und andere innerstädtische Problemzonen angekündigt[15]. Diese, dem Zentralstaat direkt unterstellten Urban Development Coporations sollten sich einzig und allein auf die Aufgabe der Revitalisierung klar abgegrenzter Zonen konzentrieren. Man hoffte durch solche «Single-minded»-Institutionen rascher zu Ergebnissen zu kommen. Kritisiert wurde, daß die Kommunen mit ihren diffusen Interessen und den Rücksichtnahmen auf diverse Gruppen jeweils nur Entwicklungskonzepte vertreten konnten, die auf einem breiten Konsens beruhten. Gerade dieser breite Konsens wurde in Widerspruch zu den ökonomischen Notwendigkeiten gesehen.

Aus dem Antagonismus zwischen konservativer Zentralregierung und häufig labourregierten Kommunen entwickelte sich die Vorstellung, daß die Urban Development Corporations die kommunale Planungshoheit in den ihnen übertragenen Gebieten übernehmen sollten. Dabei wurde ihnen aufgegeben, die Entwicklung nicht aufgrund abstrakter Planungskonzepte sondern aufgrund enger Kontakte zu Investoren und ihrer Investitionsplanungen zu entwickeln. Das neue Schlagwort lautete «market led development».

Die Rolle der London Docklands Development Corporation bei der Entwicklung von Canary Wharf

Mit der Gründung der London Docklands Development Corporation (LDDC) wurde entsprechend der oben skizzierten Konzeption eine staatliche Entwicklunsgesellschaft verantwortlich für die Revitalisierung eines großen Gebiets östlich der City. Kernstück sollte hier zunächst die Entwicklung von Canary Wharf auf der Isle of Dogs sein. Dieses Gebiet wurde gleichzeitig zur Enterprise Zone erklärt, d.h. Büroinvestitionen können sofort abgeschrieben werden, die Verluste können mit anderen Erträgen verrechnet werden. Gleichzeitig besteht eine zehnjährige Steuerbefreiung von den «rates», die nach Anhebung der Businessrates im Jahre 1989 einen erheblichen Vorteil mit sich bringt.

Als weitere strategische Entscheidung wurden der London Docklands Development Corporation alle öffentliche Grundstücke übertragen. Gleichzeitig erhielt sie das Recht Grundstücke zum «current use value» von privaten Eigentümern durch «compulsory purchase» an sich zu ziehen. Diese Rechte stehen im übrigen auch den Kommunen zu. Die LDDC ist hier den Kommunen gleichgestellt. Diese Bildung von Grundstückspools hat sich als wichtige strategische Entscheidung für alle ähnlichen Revitalisierungsaufgaben herausgestellt. Ohne die Verfügung über Grundstücke wären die Investitionserfolge nicht möglich gewesen.

Kooperation mit privaten Investoren «market led development»

Entsprechend des Auftrags sollte die London Docklands Development Corporation keine eigenen Entwicklungsziele oder Pläne vorlegen, die dann mit Hilfe von Subventionen

256

oder anderen Anreizen durchgesetzt werden sollten. Ihre Aufgabe war es vielmehr die Absichten von Investoren zu erkunden, gleichsam Marktforschung für die Entwicklung des Gebietes zu treiben, um gestützt auf diese Ergebnisse dann einen Revitalisierungsprozeß zu begleiten. Die Gesellschaft hat sich von Anfang an für ein großräumiges Infrastrukturkonzept verantwortlich gefühlt, insbesondere hat sie die Docklands durch ein schienengebundenes Verkehrsmittel erschlossen. In der Öffentlichkeit wird kritisiert, daß die Infrastrukturinvestitionen nicht rechtzeitig und nicht in ausreichender Kapazität bereitgestellt wurden und daß z.B. das Light Rail System der Docklands nicht in das übrige öffentliche Personennahverkehrssystem Londons integriert ist. Diese Kritik ist berechtigt. Man muß jedoch berücksichtigen, daß die London Docklands Development Corporation genauso wie die Öffentlichkeit von dem Tempo der Entwicklung nach 1984 völlig überrascht wurde. Aus der Erfahrung der 70er Jahre erschien es als die größte Sorge, überhaupt Investoren zu gewinnen. Niemand hat den Boom, der sich dann später herausstellte, in dieser Form vorhergesehen.

Vor dem Hintergrund der Sorge, Investoren zu gewinnen, wurden in der Anfangsphase praktisch alle Investitionen, die sich anboten zugelassen. Entsprechend der Philosophie der Enterprise Zones wurden keine planerische Vorgaben gemacht. Das Ergebnis der ersten Investitionswelle war dementsprechend problematisch und wird auch von der Docklands Corporation selbst inzwischen nicht mehr als akzeptabel angesehen. Es kam zu einem unkoordinierten Nebeneinander unterschiedlicher Bauformen und Nutzungsdichten. Zum Teil entstanden bizarre Gebäude, deren Architektur kaum längerfristig Bestand haben konnte. In der Zwischenzeit erweist sich die erste Generation der sehr lockeren Bebauung schon als veraltet. Man muß davon ausgehen, daß Gebäude aus der Frühphase der Entwicklung schon in wenigen Jahren wieder abgerissen werden.

Den entscheidenen Durchbruch erreichte die London Docklands Development Corporation etwa um 1984, als sich in London eine Verknappung des Büroangebotes andeutete und größere Büroinvestoren die auch flächenmäßig günstigeren Entwicklungschancen in den Docklands erkannten. Dabei waren die Baumöglichkeiten in der City aufgrund der damals geltenden Dichtevorschriften weitgehend ausgeschöpft. Die City schien am Ende ihrer Bebauungskapazitäten angekommen zu sein. Vor diesem Hintergrund bildete sich ein erstes Investorenkonsortium, das sich unter dem Namen «Canary Wharf Development Companies (CWDC)» zusammenschloß. Hierzu gehörten «First Boston Properties», «Credit Suisse» und «Morgan Stanley International». Diese Grupe engagierte das Architekturbüro «Skidmore, Owings & Merrill» (SOM) im Januar 1985 um einen Gesamtplan für das Gebiet zu entwickeln. Die Gruppe übernahm es auch die Ingenieurpläne für die Infrastruktur auszuarbeiten und Prototypen für Bürogebäude zu entwickeln. Dabei wurde von Anfang an geplant, Bürogebäude zu errichten, wie sie in Höhe und Volumen bisher nicht üblich waren. Die Canary Wharf Development Companies erhielten damit die Chance abgesehen von den großen Infrastrukturentscheidungen, die von der London Docklands Development Corporation schon gefällt waren (Schienenverkehr), ihr Gebiet weitgehend von Grund auf neu zu beplanen. Hierbei wurden Bebauungsplanung und Gebäudeplanung in einer Hand zusammengefaßt. Die London Docklands Development Corporation geriet demgegenüber eher in die Rolle eines Schiedsrichters, der von den privaten Investoren entwickelte Pläne begutachtete und korrigierte. Hier wird ein völlig anderes Planungs- und Kooperationsverständnis zwischen öffentlicher Planungsbehörde und privaten Investoren deutlich. Die LDDC hat zunächst keine eigenen Entwicklungsvorstellungen vorgelegt, sondern startete einen Diskussions- und Verhandlungsprozeß, in dem die Privaten ihrerseits ihre Investitionsabsichten artikulierten. Dabei hat sich im Laufe der Zeit eine Gewichtsverlagerung ergeben. Während in der ersten Phase das Konzept «Market led Development» noch wörtlich genommen wurde, übernahm die Docklands Corporation in der Folgezeit eher die Funktion eines Moderators, der Standortpotentiale analysiert, sie mit den beabsichtigten Investitionen vergleicht und unterschiedliche Investitionskonzepte verschiedener Investoren koordiniert. Neben der Verlagerung der Gewichte bei der Entwicklung des Gebietes zugunsten privater Investoren, erreichte das Gesamtprojekt Dimensionen, wie sie bisher auch in London nicht aufgetreten waren. Nach Fertigstellung soll in Canary Wharf eine Größenordnung bis zu einer Million Quadratmeter Büroflächen entstehen – sowie eine Größenordnung von 50.000 Quadratmeter Einzelhandelsflächen, Restaurants und Freizeitflächen. Es sind mehrere Hotels geplant. Das größte Gebäude erreicht eine Höhe von 50 Stockwerken. Selbst die leistungsstarke Canary Wharf Development Gruppe, die sich sehr früh gebildet hatte, konnte diese Dimension nicht alleine realisieren. So hat seit 1987 die kanadische Gesellschaft Olympia & York die Führung bei der Entwicklung von Canary Wharf übernommen. Dabei ist zu berücksichtigen, daß die privaten Entwicklungsgesellschaften nicht nur die Bürotürme errichten, sondern gleichzeitig die Tiefgaragen- und Straßeninfrastruktur und die sonstige Entsorgungs- und Versorgungsinfrastruktur finanzieren und in eigener Regie entwickeln.

Ein neuer Typus Entwicklungsplanung?

Gegen ihr eigenes anfängliches Selbstverständnis hat die LDDC relativ früh in Auseinandersetzungen mit den Investoren eigene Entwicklungsvorstellungen, die sie dann auch gegenüber den Investoren durchsetzte, formuliert. Durch ein 'learning by doing' hat sich in einer offenen, durch rechtliche Vorgaben kaum fixierten Situation folgende Rolle der LDDC herausgestellt:
– Die LDDC definierte grobe Nutzungskonzepte, wie sie z.B. in Flächennutzungsplänen festgelegt werden.
– Diese groben Festlegungen wurden veröffentlicht. Dabei hat die LDDC zum Teil Architekten beauftragt, mögliche Lösungen zeichnerisch in Skizzenform zu fixieren. Diese Projektskizzen wurden wiederum von den Investoren zum Anlaß für eigene Weiterentwicklungen genommen. Gestützt auf solche groben Bebauungs- und Nutzungskonzepte wurden dann Details der Vertragsverhandlungen über die Grundstücksveräußerung geregelt.
– Durch vertragliche Auflagen, nicht durch öffentlich rechtliche Pläne wurden im Zusammenhang mit dem Verkauf Art und Inhalt der baulichen Nutzung festgeschrieben. Der Zuschlag bei der Veräußerung erfolgte einmal nach der Höhe des Gebots, daneben aber auch nach der Qualität und der Art der geplanten Investitionen und Nutzungen. Dabei hat die LDDC implizit auch Bewertungen über die Qualität der Investitioen und über die Komplementarität verschiedener Investitionsprojekte angestellt. Sie wurde im Ergebnis durch ihre Marktmacht und die Einflußmöglichkeiten, die sie als Grundstückseigentümerin hatte zum globalen «Schiedsrichter» über die Entwicklung der Region, wobei ihre Vorstellungen über die

Ausnutzung der Standortpotentiale und die Nutzungsmischungen bzw. über komplementäre Investitionen eine erhebliche Bedeutung erhielten.

- Im Wohnungssektor hat die LDDC Grundstücke an Bauträger zum Teil nur im Erbbaurecht vergeben, um Einfluß auf Verkaufspreise und Art der Bebauung nehmen zu können.
- Aus der Praxis der Grundstücksentwicklung hat die LDDC damit eine öffentliche Planungsrolle entwickelt, die gestützt auf die Eigentumsrechte und die Einflüsse bei der Bereitstellung der Infrastruktur sehr detailliert sein konnte. Aufgrund der engen Kooperation wurden die Investitionsplanungen der privaten Investoren und die öffentlichen Planungen praktisch in einem Prozeß integriert. Diese Integration zweier ansonsten getrennter Planungsvorgänge hat einen Zeitgewinn gebracht und darüber hinaus auch Reibungsverluste, wie sie bei der Anpassung privater Planungen an vorgegebene öffentliche Planungen entstehen können, vermieden. Allerdings hat sich im Ergebnis der Einfluß der Investoren auf die Gestaltung und Bebauung der Gebiete erheblich vergrößert. Wie man heute im Nachhinein feststellen kann, fehlt es an «Checks und Balances». Die Investorenabsichten wurden in ihren Inhalten von der LDDC zu wenig hinterfragt. Dies erklärt die enormen Baumassen und Bauvolumen, die etwa in Canary Wharf errichtet wurden. Andere Investoren verweisen darauf, daß in Gebieten mit stärkerem Einfluß der Kommunen und der umgebenden Nachbarschaften Investitionen und Gebäude kaum hätten realisiert werden können. Inzwischen zeigt sich, daß auch die Nachfrager diese extremen Baumassen nicht als optimal ansehen.

Exkurs: Zur Veräußerungspraxis der LDDC

Die LDDC war durch politische Vorgaben verpflichtet, Grundstücke weitgehend zu veräußern. Selbst private Entwicklungsgesellschaften verweisen darauf, daß diese Strategie suboptimal war. Die Verkaufspreise zu den jeweiligen Zeitpunkten spiegelten die künftigen Wertsteigerungen nicht wider. Es kam zu spekulativen Weiterveräußerungen. Nach Auffassung vieler Experten wäre es sinnvoller gewesen, Grundstücke nicht zu veräußern und auf Dauer im Eigentum einer Entwicklungsgesellschaft zu halten. Dafür hätten Joint-Venture-Strategien oder auch Erbbaurechtslösungen gewählt werden können. Diese Praxis wird von privaten Eigentümern und auch von der City of London stärker angewandt. Die Veräußerungsauflage wurde auch von Vertretern der LDDC z.T. als rein politische Vorgabe angesehen, die wirtschaftlich nicht optimal sei. Hintergrund dieser «Veräußerungseile» scheint zu sein, daß die konservative Regierung möglichst weitgehend vollendete Tatsachen schaffen wollte, um zu verhindern, daß unter anderen politischen Vorzeichen Entwicklungen wieder in andere Richtungen gesteuert werden können.

Auseinandersetzung mit Kritik

In der öffentlichen Diskussion, vor allem in der Fachpresse, wurde die Entwicklung in Canary Wharf weitgehend kritisch kommentiert. Stellvertretend für vieles sei hier ein Zitat aus «Angewandte Geographie»[16] herangezogen:

«Stadtplaner und Politiker aller Herren Länder sind in den vergangenen Jahren nach London gepilgert, um Margret Thatchers Prestigeprojekt «Docklands» zu bewundern.

Das Patentrezept, das ihnen für die Sanierung innerstädtischer Problemgebiete in die Hand gedrückt wurde, lautete: Bilde eine «Urban Development Corporation», statte sie mit den erforderlichen Kompetenzen aus und überlasse den Rest dem freien Markt. Wenn es jedoch nach einem im Juni erschienenen Bericht des Docklands Consultitive Committee (DCC) geht, so basiert dieses auf reinem Wunschdenken – nach Ansicht des DCC sind die Docklands nämlich heute vor allem prädestiniert, die ganze Spannbreite möglicher Fehler bei der Durchführung eines innerstädtischen Umstrukturierngs- und Entwicklungsprojektes zu demonstrieren.»

Versucht man die Kritik zu sortieren, dann lassen sich folgende Argumente heraus-destilieren:

«From Boom to Bust»:
- Mit diesem Schlagwort wird die überzogene Investitionstätigkeit, die jetzt zu hohen Leerständen führen wird, kritisiert. Hierbei ist auf die Ursachen der Über-investitionen zu verweisen (vergleiche das Kapitel über die Erklärung der Zyklen). Den neugewählten Planungs- und Steuerungsverfahren kann dieser extreme Zyklus nur zu einem winzigen Teil angelastet werden. Die Ursachen liegen vielmehr in der Konkurrenz zwischen den Dockland und der City und in den überzogenen allge-meinen wirtschaftlichen Wachstumserwartungen, die sich als Folge der notwendig werdenden Stabilisierungspolitik seit 1988/89 nicht bestätigten. Im Development of the Environment (DOE) wird die Überinvestition allerdings als weniger problematisch angesehen. Es wird darauf verwiesen, daß die Investitonen im großen Stil von ausländischen Investoren finanziert wurden und daß London durch das Überangebot über Jahre hinaus zu einem attraktiven Bürostandort geworden ist. Die Wettbewerbs-fähigkeit Londons als Bürostandort sei durch den Bauboom erheblich gesteigert worden.

Übersubventionierung:
- Diese Kritik scheint aus der heutigen Sicht weitgehend berechtigt. Die Sofort-abschreibung, massive Kapitalzufuhr für Infrastruktur und andere öffentliche Leistungen haben Investoren dazu veranlaßt, innerhalb kurzer Fristen extrem hohe Investitionsvolumen zu realisieren. Auch die Größe der Gebäude und das Gesamtvolumen sind zum Teil Folge dieser Investitionen. Allerdings muß man auf der anderen Seite berücksichtigen, daß die Tertiärisierung der Wirtschaft voranschreitet und auch im industriellen Sektor immer mehr Beschäftigte in Büros arbeiten. Immer mehr Büros separieren sich von den Produktionsanlagen. Dement-sprechend muß der Staat bzw. die Kommunen die Voraussetzungen für solche Investitionstätigkeit schaffen. In der Bundesrepublik stehen alle Großstädte von Frankfurt bis München vor der Aufgabe, neue Bürobauten in verkehrsgünstigen Lagen zuzulassen (vergleiche die Diskussion um die Umwidmung der Speicherstadt und der angrenzenden Gebiete in Hamburg, die Diskussion in Frankfurt über die Genehmigung weiterer Hochhausbauten, oder die Diskussion in München über die Konzentration von Bürotätigkeit am mittleren Ring).

Der Mythos des «Trickle-down»:
- Mit dieser Kritik wird darauf verwiesen, daß trotz der riesigen Investitionen in den Docklands zu wenig Arbeitsplätze für die Bevölkerung in den umliegenden Boroughs

entstehen. Die Beschreibung ist grundsätzlich richtig. Unabhängig davon bleibt es wahrscheinlich ein Faktum, daß eine Reindustrialisierung in der erforderlichen Größenordnung unmöglich war. Die Veränderung der Qualifikationsstrukturen einer Stadtregion, die zum einem wichtigen Dienstleistungszentrum der ganzen westlichen Welt geworden ist, war rascher als die Anpassungsmöglichkeiten der Qualifikationsstruktur der vorhandenen Arbeitskräfte. Damit kommt es zu einer ständigen Zuwanderung von Spezialisten von außerhalb Londons. Trotz steigender Zahl der Arbeitsplätze und steigender Arbeitsnachfrage bleibt eine strukturelle Arbeitslosigkeit bestehen. Dieses Dilemma läßt sich nur durch sehr komplexe Strategien lösen. Dabei liegt auf der Hand, daß hierfür eine Kombination aus Umschulung, Fortbildung und der Förderung von einfachen Arbeitsplätzen notwendig gewesen wäre. Es kann nicht Aufgabe privater Investoren sein, diese Rolle zu übernehmen. Insofern läuft die Kritik ins Leere. Sie müßte sich an die gesamte Bildungs- und Strukturpolitik der Regierung wenden. Eine Stadtentwicklungsgesellschaft vom Typ «Docklands» kann nicht die Verantwortung für das Erreichen der Vollbeschäftigung in ihrem Bezugsbereich übernehmen. Zur Lösung der Arbeitslosenprobleme wäre in Großbritannien Strategien erforderlich gewesen, wie sie z.B. in der Arbeitsmarktpolitik in Schweden entwickelt wurden. Der LDDC als eine spezielle, stark auf die Grundstücksentwiclung konzentrierte Organisation kann man diese Arbeitslosigkeit nicht anlasten.

Ungelöste Transportprobleme
– Die Hinweise auf die bis heute nicht befriedigend gelösten Transportprobleme sind berechtigt. Allerdings bleibt erstaunlich, wie es den privaten Investoren gelungen ist, für den Baustellenverkehr ein Transportsystem über die Themse zu organisieren. Die unzureichende Parallelität von privaten Investitionen und öffentlichen Verkehrsinvestitionen geht sehr stark auf die Unterschätzung der Investitionsmöglichkeit zurück. Koordinierungsfehler dieser Art lassen sich aus der Retroperspektive jeweils leicht brandmarken. Man muß akzeptieren, daß in der Ex-ante-Perspektive niemand den tatsächlich eingetretenen Bauboom erwartete und die Infrastrukturplanungen dann in den erforderlichen Zeiten nicht mehr umgesetzt werden konnten.

Verschärfung des Wohnungsmangels
– Durch das Angebot von rund 50.000 zusätzlichen Arbeitsplätzen im Bereich hochwertiger Dienstleistungen ist die Nachfrage nach Wohnungen erheblich gestiegen. Gleichzeitig hat der Wohnungsmangel bei der alteingesessenen Bevölkerung zugenommen. Die Wohnungsbautätigkeit, die die London Docklands Development Corporation angeregt hat, richtet sich überwiegend auf die Bedürfnisse der neuen Nachfrager. Sozialer Wohnungsbau wurde nur im begrenzten Umfang finanziert. Letztenendes richtet sich die Kritik an die Zentralregierung, da die Kommunen und die Docklands Corporation nicht die Möglichkeit haben sozialen Wohnungsbau im großen Stil zu finanzieren. Am Beispiel der Docklands wird besonders deutlich, welche Verteilungskonflikte entstehen, wenn eine kräftige Marktentwicklung zu Nachfragesteigerungen im Wohnungssektor führt und das Angebot nur aus privaten, frei finanzierten Wohnungen besteht. Zusammenfassend kritisiert das Docklands Consultative Committee in einer Studie:[17]
«It is now widely recognised that there is a «Docklands Factor» which although not limited to London Docklands, is one of the distinctive characteristics of the redevelop-

ment of this area. It is made up following: waterside «luxury flats» and up-market office and retail development; polarisation between the existing and new populations; few benefits for existing residents; and lack of consultation with local communities.

Most of the current concerns about Docklands development are not new. Local people and local authorities warned from the time the LDDC was set up in July 1981, that if planning was ignored and market forces allowed to take over, those in need of affordable housing, the disadvantaged, or those seeking jobs and job training would lose out. It was also predicted that local people would not be properly consulted and that existing industrial firms would be driven out. All these things have happened; to a greater degree even than was feared.

More recently, others have joined a growing number of critics. First, businesses and developers, the very sectors that supported setting up the LDDC and an general relaxation of planning controls, are now demanding «strategic planning» and in particular forward planning for transport infrastructure. The chorus of disapproval over transport provision recently forced the Government and the LDDC into an intensive period of public relations activity aimed at reassuring the business community.

Professional planning and architectural observers have attacked the lack of planning in Docklands and its incoherent and chaotic urban from, particularly on the Isle of Dogs.»

Folgerungen für die Bundesrepublik

Angesichts dieser Kritik muß man sich mit der Frage auseinandersetzen, welche Folgerungen für die Bundesrepublik zu ziehen sind.

Dabei bleibt natürlich darauf zu verweisen, daß Investitionsvolumina, wie sie in den Docklands zu bewältigen waren, in der Bundesrepublik außerhalb Berlins kaum auftreten.[18] Dennoch gibt es auch in der Bundesrepublik immer häufiger größere innerstädtische Entwicklungsprojekte. Beispiele:
- Hamburg: Speicherstadt und angrenzende Gebiete;
- Düsseldorf: Hafenentwicklung;
- Duisburg: Hafenentwicklung;
- Frankfurt: Hafenentwicklung;
- Berlin: Potsdamer Platz, Otkreuz, Westhafen, Halensee und verschiedeneandere Projekte;
- Berlin: Hauptstadtbauten.

Auch in diesen Gebieten müssen innerstädtische Grundstücke, die zum großen Teil der öffentlichen Hand gehören, innerhalb kurzer Fristen neu bebaut werden. Die Entwicklung am Potsdamer Platz wird z.B. zu einem größeren Bauvolumen führen als in Canary Wharf. Bei solchen innerstädtischen Großvorhaben erscheinen uns folgende Erfahrungen übertragbar:
- Die Bildung von Grundstücks-Pools in der Hand einer öffentlichen Entwicklungsgesellschaft erscheint als Voraussetzung sinnvoll, wenn nicht sogar notwendig.[19]
- Anders als in den Docklands sollte eine solche Entwicklungsorganisation jedoch in der Verantwortung der Kommune gegründet werden, dabei können sich durchaus private Investoren beteiligen. Auch eine Landesbeteiligung erscheint sinnvoll. Die Mediaparkgesellschaft in Köln erscheint hier als ein gutes Beispiel für die

Organisation solcher Entwicklungsaufgaben.
- Die Ziele der Urban Development Corporations jeweils über eine «Single-Minded» Organisation zu verfügen, die sich bestimmten Planungsaufgaben voll widmet, erscheint aufgrund der englischen Erfahrungen als erfolgreich bestätigt. Grundstücksbezogene, zeitlich befristete Entwicklungsorganisationen als Hauptanlaufstelle werden den komplexen Aufgaben besser gerecht als eine Vielzahl kommunaler Ämter, die jeweils noch mit zahlreichen anderen Aufgaben belastet sind. Allerdings ist es in der Bundesrepublik allein aus rechtlichen Gründen unmöglich, die kommunalen Kompetenzen zugunsten einer staatlichen Entwicklungsorganisation soweit zurückzudrängen wie in Großbritannien.
- Es empfiehlt sich, frühzeitig mit potentiellen Investoren Kontakt zu halten und eine Planung durch «Bargaining» mehr aus einer katalytischen Einstellung heraus zu realisieren. Bei komplexen, zusammenhängenden Erneuerungsaufgaben oder bei anderen Großprojekten lassen sich Planungen nicht von außen aufoktroyieren. Im Vergleich zu den Planungen in Kings Cross und den Docklands zeigt sich allerdings, daß ein starker, an den Interessen der Öffentlichkeit orientierter Planungsprozeß zu einer besseren Interessenabwägung führt. Auch in Kings Cross haben private Architekturbüros zusammen mit möglichen Investoren ein Bebauungs- und Nutzungskonzept entwickelt und dies der Öffentlichkeit präsentiert. Im Zuge der öffentlichen Diskussion mußten diese Pläne mehrmals revidiert und angepaßt werden. Entscheidend war ein System von «Checks and Ballances» zwischen privaten Interessenten und Investoren sowie den beteiligten Kommunen. In den Docklands wurden die kommunalen Interessen zu weit zurückgedrängt. Dies führt zu dem Ergebnis, daß die Investoren ein zu «freies Feld» vorfanden, was im Nachhinein auch ihren eigenen Interessen nicht gedient hat. Eine Balance der Einflüsse und Interessen, d.h. die Balance zwischen privaten und öffentlichen Interessen ist von Anfang an erforderlich. Dabei erscheint es nicht entscheidend, daß die öffentliche Hand zunächst Bebauungspläne vorlegt, die dann von Privaten «ausgefüllt» werden. Es kann durchaus sinnvoll sein, daß Bebauungskonzepte und Bebauungsplankonzepte in einer Hand in einem Wurf formuliert werden. Allerdings muß immer die Möglichkeit bestehen, daß unabhängig von der technischen Verantwortung für diese Planungen öffentliche Interessen berücksichtigt und notfalls nachdrücklich eingebracht werden können. Politisch selbstbewußte Kommunen werden in jedem Fall ihre Planungsmacht auch in Marktmacht umsetzen können, d.h. sie werden im Verhandlungswege die öffentlichen Interessen wahrnehmen können, auch wenn sie die technische Kompetenz zur Formulierung und Aufstellung der Pläne auf private Investoren übertragen haben.
- Private Investoren tendieren dazu, Architektur und große Bavorhaben als Teil einer «Corporate Ego-Darstellung» anzusehen. Die in der Bundesrepublik vor allem in Frankfurt zu beobachtende «Corporate Ego Konkurrenz» stärkt die Tendenz, jeweils möglichst ausgefallene und individualistische Bauformen und Gebäude zu entwickeln. Dies fördert eine Heterogenität und eine Vielfalt des Erscheinungsbilds. Dadurch wird die Attraktivität einer Stadt kaum erhöht. Noch entspricht eine solche Lösung ihrem Selbstverständnis. Das Beispiel von Canary Wharf zeigt, daß dann, wenn man privaten Investoren freies Feld bei der Planung und dem Bau von großen Büroprojekten verschafft, die «Corporate Ego Komponente» einer machtvollen Selbstdarstellung über die Nutzerinteressen triumphieren kann. In der Bundesrepublik müssen sich die Kommunen in Zukunft stärker auf dieses Element des

Selbstdarstellung von Großunternehmen durch Architektur und Großbauten einstellen. Die Kommunen tun gut daran, sehr frühzeitig klare Positionen zu beziehen, damit ihre Städte nicht zum Austragungsort des Selbstdarstellungswettbewerbs von Großunternehmen und Stararchitekten werden, die dabei jeweils eine enge Symbiose eingehen.

In Canary Wharf ist die Balance zu stark zugunsten privater Großinvestoren verschoben worden. Dies erscheint als struktureller Fehler, der von Anfang an bestand. Dies wird auch deutlich im Vergleich zu den Projekten Broadgate und Kings Cross.

Insgesamt zeigt das Experiment mit der London Docklands Development Corporation, daß eine rein «market led development» nicht möglich ist. Ohne daß dies von der London Dockland Development Corporation so gewollt und beabsichtigt war, ist sie in eine planende Rolle, die grobe Nutzungsstrukturen und ein städtebauliches Grundmuster vorgibt, hineingewachsen. Sie hat allerdings Projektentwicklung und die Formulierung eines Bebauungsplans de facto zu einer integrierten Planungsphase zusammengefaßt. Dies hat zu einer erheblichen Beschleunigung der Abwicklung geführt. Darüber hinaus war die enge Kooperation zwischen der LDDC und den wenigen Investoren sehr erfolgreich bei der Koordinierung sehr komplexer Aufgaben.

Daraus ergibt sich insgesamt die Folgerung, daß etwa bei der Planung und Vorbereitung der Hauptstadtbauten, bei komplexen Recyclingaufgaben (z.B. Entwicklung des Hamburger oder Düsseldorfer Hafens) eigenständige Entwicklungsorganisationen mit einer großen Autonomie durchaus sinnvoll sind. Die zeitliche und sachliche Integration der Projektentwicklungen mit der Planung ist bei hochkomplexen Investitionen mit hohem Volumen sinnvoll. Mit großer Befriedigung kann man aus der Sicht der Beteiligungsregeln des Baugesetzbuches zur Kenntnis nehmen, daß das Zurückdrängen des kommunalen Einflusses und das Zurückdrängen der Bürgerbeteiligung zwar das Realisierungstempo der Investitionsprojekte erhöht hat, nicht jedoch ihre Qualität. Immer wieder bestätigt sich die Erfahrung, daß Städtebau ohne Kontrollen, d.h. Städtebau unter dem Gesichtspunkt enger Interessen, seien sie Interessen der Kommunen oder von Investoren, leicht dazu tendiert, einseitig zu werden und in seiner Qualität hinter ausdiskutierten Lösungen erheblich zurückbleiben kann. Man kann aus dieser Aussage zwar nicht folgern, daß Beteiligung Qualität garantiert. Mit hoher Wahrscheinlichkeit gilt jedoch, daß Städtebau auf «freiem Feld» (im doppelten Sinn des Wortes) durchaus dazu tendiert, unbalancierte Lösungen zu entwickeln, die langfristig nicht als optimal gelten können.

Die Entwicklung von Broadgate durch Rosehaugh Stanhope

Broadgate – Eine Entwicklung am Cityrand durch private Investoren

Die Entwicklung von Broadgate stellt von der Konzeption wie von der Lage her einen Kontrast zu Canary Wharf dar. Durch Broadgate wird die Liverpool Street Station mit einem großen Büro- und Einzelhandelskomplex eingefaßt, zum Teil werden dabei der Bahnhof und die Gleise überbaut. Liverpool Street Station liegt als typischer Kopfbahnhof am nordöstlichen Rand der City. Die angrenzenden Wohngebiete gehören nicht zu den attraktivsten Londons. In der Vergangenheit bildeten sich um den Bahnhof herum ein Kran von wirtschaftlich nicht sehr starken Nutzungen (alte Geschäfte, Lagerfunktionen, einige Büros, Wohngebäude niedriger Attraktivität, zum Teil Stadtbrache). Rosehaugh Stanhope haben diesen Standort bewußt ausgewählt, weil sie zu Beginn der 80er Jahre von der Analyse ausgingen, daß innerhalb der bisherigen Citygrenzen nicht genügend Büroraum angeboten werden konnte. Außerdem war abzusehen, daß die Verkehrsprobleme der City ständig wachsen. Die unmittelbare Nähe zu einem Bahnhof erschien deshalb als ein wichtiger Standortfaktor. Die Standortwahl erschien aus der Sicht der ersten Hälfte der 80er Jahre ausgesprochen riskant. Ein Projekt von rund 300.000m² Büro- und Einzelhandelsflächen in einer wenig prestigeträchtigen Lage, unmittelbar an den einfachen Wohngebieten Ost-Londons zu planen und zu finanzieren, stellte eine ausgesprochene Pionierleistung dar. In der Zwischenzeit hat sich gezeigt, daß durch die hohe Qualität der Architektur und der stadträumlichen Gestaltung sowie durch die Nutzungskombination eine «neue Adresse» entstanden ist. Stanhope ist es gelungen, die problematische Qualität des Standortes durch die eigene Entwicklung erheblich aufzuwerten und das selbstgesetzte Ziel «to create places» zu realisieren. Der Erfolg des Projekts wird allein deutlich, daß gegenwärtig praktisch keine Leerstände bestehen. Trotz beginnenden Überangebots war Broadgate noch voll vermietbar. Dabei wurden relativ hohe Mieten erzielt.

Zur Konzeption

Zur Vorbereitung wurden für Broadgate sehr detaillierte Marktanalysen durchgeführt. So wurde genau ermittelt, welche Sektoren in den kommenden Jahren überdurchschnittlich rasch wachsen würden. Es wurde sehr frühzeitig mit potentiellen Nutzern Kontakt aufgenommen, um deren Raumbedürfnisse zu analysieren. Stanhope hat es inzwischen zu einem Markenzeichen seiner Konzeptionen gemacht, Projektentwicklung jeweils auf sehr detaillierte Marktforschung über Nutzerbedürfnisse und aufgrund von Erfahrungen mit ähnlichen Projekten durchzuführen. Bei den Vorplanungen stellte sich heraus, daß Broadgate auf die Bedürfnisse der Finanzdienstleister aber auch auf dazu komplementären Branchen (support services) hin orientiert sein sollte. Um die Raumansprüche der potentiellen Kunden genau zu erfassen, wurde ein zum Teil erheblicher Aufwand betrieben. So wurden Nutzer- und Marktstudien in den USA durchgeführt, um festzustellen, welche Arbeitsabläufe in «Trading Floors» tatsächlich zu beobachten sind und welche Größen bzw. welche Büroformen diesen Arbeitsabläufen am besten entsprechen. Es wurden Arbeitsgemeinschaften mit den möglichen Nutzern gebildet, um deren Raumansprüche aus ihren Arbeitsabläufen heraus zu analysieren.

Bei der Planung machte man sich die in den 80er Jahren größeren Freiheiten beim Bürobau zunutze. In der Masse wurden sehr tiefe, kompakte Gebäude geplant. Auf einer Etage gibt es bis zu 5.000 Quadratmeter Nutzfläche. In diesen tiefen Gebäuden wurden effektive Ausnutzungen erreicht, die in Hochhäusern nicht möglich sind. Auf diese Weise wurden sehr günstige Relationen zwischen Brutto- und Nettogeschoßfläche erzielt. Architektonisch ging man von der Vorstellung aus, daß sich der neue Bürokomplex möglichst wenig störend in die bestehende Umgebung einfügen sollte. Dies wurde einmal durch die Gleisüberbauung und die enge Umrandung des Bahnhofs erreicht. Dadurch mußten nicht große neue Flächen freigelegt werden. Durch eine geschickte architektonische Gestaltung wurde versucht die neun- bis elfstöckigen Bauweisen nicht als Türme sondern als große kompakte, eher «tankerartige» Gebäude zu errichten. In der Tradition der Londoner Stadtgrundrisse bleibend, wurden innerhalb des Bürokomplex kleine Plätze angelegt. Durch eine geschickte Plazierung der Einzelhandelsgeschäfte sollen Fußgänger möglichst an belebten Erdgeschoßfassaden mit Schaufenstern und Auslagen und nicht an toten Bürozonen vorbeigehen. Dort wo sich auch in der Erdgeschoßzone Büros befinden, wurden diese sehr offen zum Außenraum hin angelegt, d.h. man hat einen Blick in dieBüros. Dies war allerdings nur möglich, weil Broadgate einen in sich abgeschlossenen Komplex bildet und von daher große Fußgängerströme in weiten Bereichen während der Rush-Hour auftreten.

Beziehungen zur City und den angrenzenden Boroughs

Bei der Entwicklung von Broadgate wurde versucht, nicht nur den Kernbereich der Finanzdienstleistungen als Kunden anzusprechen, sondern darüberhinaus die sog. «Support Services» ebenfalls für Broadgate zu interessieren. Der Bereich der «Support Services» reicht von Werbung bis Druckereifunktionen, die heute überwiegend an Computerarbeitsplätzen erfüllt werden, bis hin zu Wirtschaftsprüfungs- und Makler- organisationen, die city-bezogen agieren. Durch dieses breite Spektrum von unterschied- lichen Arbeitsplätzen sollte auch die Zustimmung der angrenzenden Boroughs erreicht, die daran interessiert waren, nicht nur hochspezialisierte Finanzdienstleistungs- Arbeitsplätze in ihren Gebieten zu haben.

Grundstückssituation

Die Grundstücke gehörten vor der Neuentwicklung weitgehend British Rail bzw. der Grundstücksgesellschaft von British Rail. Stanhope hat diese Grundstücke erworben. Dabei wurden entwicklungsorientierte Kaufpreise vereinbart, d.h. Rosehaugh Stanhope erhielten die Grundstücke nicht zu einem Festpreis. Es wurden viel mehr Zusatzzahlungen für den Fall vereinbart, daß die Mieten bestimmte Niveaus überschreiten würden. British Rail hat auf diese Weise die Entwicklung erleichtert und hat durch die Vertragsklauseln sichergestellt, daß dem Veräußerer spätere Ertragssteigerungen ebenfalls noch zugute kommen.

Projektplanung und Realisierung

Das Projekt mußte unter schwierigen Bedingungen realisiert werden. Der Bahnhofsverkehr durfte wegen der Bedeutung für die City nicht unterbrochen werden. Dies hatte zur Folge, daß in kritischen Phasen Schlüsselinvestitionen nur während weniger Stunden in der Nacht realisiert werden konnten. Angesichts der Innenstadtlage sollte die Bauzeit möglichst kurz bleiben. Dies wurde erreicht, in dem die Gebäude im Detail an anderer Stelle vorgeplant und z.T. vorgefertigt wurden. Am Bauplatz selbst mußten sie dann nur noch «zusammengesetzt» werden.

Bewertung

Stanhope verweisen darauf, daß es gelungen sei, in hochwertiger Cityrandlage hohe Dichten zu realisieren, ohne die Störungen hervorzurufen, die durch Hochhäuser entstehen. Tatsächlich ist es gelungen durch die kompakte Bauweise, die Bahnhofsüberbauungen und die «Durchwegung» des Gesamtprojekts trotz der riesigen Baumassen eine sehr durchlässige Bürozone zu errichten. Der Anteil der Einzelhandelsflächen liegt bei gut 10%. Angesichts der Gebäudetiefen führt dies rein optisch dazu, daß die nach außen orientierten Bereiche zu einem erheblichen Teil von Geschäften genutzt werden.

Für die Bundesrepublik bleibt offen, ob die entstehenden Büroformen (z.T. Großraumbüros) von deutschen Nutzern akzeptiert werden. Es bleibt allerdings darauf zu verweisen, daß durch Lichthöfe und geschickte Architektur immer wieder versucht wurde, eine Auflockerung der kompakten Baumassen zu erzielen.

Durch die Kooperation mit mehreren Architekturbüros und die Aufteilung des Gesamtkomplexes in zahlreiche Einzelgebäude wurde der Eindruck vermieden, daß hier ein homogenes Projekt nach einheitlichem Schema entwickelt wurde. Dennoch bleibt es natürlich ein Faktum, daß die Investitionen innerhalb kurzer Frist realisiert wurden und von daher in ihrer gesamten Erscheinungsform und Konzeption dem Verständnis dieser Bauperiode entsprechen.

Im Vergleich zu Canary Wharf ist die Broadgate-Architektur weniger monumental und erdrückend. Die größere Feingliedrigkeit der Gebäude erzeugten einen größeren Erlebnisreichtum. Durch die günstige Lage zur Bahn werden große Verkehrsmengen in kurzen Fristen bewältigt. Trotz dieser geglückten Lösung bleibt natürlich ein Projekt dieser Dimension, das in bestehende Stadtteile hineingepflanzt wird ein problematisches

Investitionsvorhaben.[20] Wenn irgend möglich, sollten solche Dimensionen vermieden werden. Auf der anderen Seite erlaubte die Großenordnung Lösungen (Gleisüberbauungen, Aufwertung eines ganzen Gebietes, hohe Komplementarität der Nutzungen), die bei kleineren Maßnahmen nicht möglich gewesen wären. Angesichts der Vorbelastung des Gebietes wären sehr viel kleinere Investitionen mit dem hohen Risiko verbunden gewesen, daß sie von den negativen Ausstrahlungen des Umfeldes stärker beeinträchtigt worden wären. Als Fazit bleibt, daß ein im Prinzip wegen seiner Größe problematisches Investitionsvorhaben mit erstaunlichem Geschick und Einfühlungsvermögen in einen bestehenden Stadtbereich eingefügt wurde. Diese Integrationsleistung war möglich, weil ein realistisches Gesamtkonzept von einem Großinvestor durchgeführt wurde. Auch hier hat die Kommune dem Investor einen großen Einfluß auf das Gesamtprojekt gegeben. Die Bebauungsplansituation erlaubte eine hohe Dichte. Im Detail waren jedoch kaum Vorgaben vorhanden, so daß der Investor einen eigenen detaillierten Plan entwickeln konnte, der von den Beteiligten Boroughs akzeptiert wurde. Die Wünsche der Boroughs in Richtung auf einen hohen Einzelhandelsanteil wurden von dem Investor aus Eigeninteresse weitgehend berücksichtigt, weil dem Investor auch daran gelegen war, nicht nur eine reine Bürozone zu enwickeln. Inzwischen zeigt sich, daß aufgrund des hohen Fußgängerverkehrs durch die Nähe zum Bahnhof die Einzelhandelsgeschäfte, auch kommerziell ein Erfolg sind.

Enwicklungen dieser Größenordnung können an mehreren Stellen in Berlin oder auch in Hamburg bzw. Düsseldorf entstehen. Hier können die Erfahrungen, wie sie bei den Investitionen in Broadgate gesammelt wurden, durchaus Berücksichtigung finden. Dies gilt sowohl für die organisatorischen Lösungen, wie für die inhaltlichen Konzepte.

Kings Cross[21]

Das Gelände liegt am Nordrand der City, am Rande zweier bestehender Bahnhöfe. Es ist weitgehend ungenutzt. Früher befand sich auf dem Gelände ein großes Krankenhaus (19. Jahrhundert). Später wurden kamen Lagerei und verschiedene Produktionsanlagen hinzu, die inzwischen stillgelegt sind. Wie in den Docklands, gab es in den letzten 20 Jahren immer wieder Erneuerungversuche. In der Zwischenzeit wurde das Gelände mehreren Transportaktivitäten überlassen. Dabei spielte der Zugang zum Kanal (Regents Kanal) eine Rolle.

Das Gelände erfreute sich eines größeren Interesses als Planungen bekannt wurden, daß dort die Endstation der Tunneleisenbahn sein sollte. Von hier führen zwei große Fernbahnen nach Norden und Nordwesten. Außerdem bestehen Anschlüsse an das Verkehrsnetz Londons. Vor diesem Hintergrund erstaunt es nicht, daß der «British Rail Property Board» zu Beginn der 80 Jahre seine Bemühungen um eine Revitalisierung verstärkte. British Rail schrieb einen Developer-Wettbewerb aus. Daran beteiligten sich Rosehaugh Stanhope mit dem Büro Norman Forster und eine andere Gruppe mit dem Büro SOM. Den Zuschlag erhielt Rosehaugh Stanhope mit Norman Forster. Der Developer-Wettbewerb wurde weniger mit dem Ziel gestartet, schon eine abschließenden städtebaulichen Entwurf vorzulegen sondern war von British Rail offensichtlich dazu gedacht, in Auseinandersetzungen mit dem Developer und seinem Architektenteam herauszufinden, welche Gruppe sich für die künftige Entwicklung am besten eignete.

Beziehungen zu den Kommunen (Boroughs)

Die Boroughs haben die konzeptionellen Bemühungen von British Rail und Rosehaugh Stanhope zunächst beobachtend zur Kenntnis genommen. Das Gelände liegt im Borough Camden. Camden hat nach der Veröffentlichung der ersten Grobkonzepte durch die Investoren sehr zähe Verhandlungen mit Stanhope begonnen. Stanhope ist sehr intensiv auf diese Verhandlungen «eingestiegen». Nach eigenen Aussagen betrugen die Entwicklungskosten bisher 5 Mio. Pfund. Diese Kosten sind u.a. sehr stark dadurch zu erklären, daß nach den ersten Grobentwürfen eine Phase längerer Anpassungen und Änderungen in Auseinandersetzung mit Camden erfolgte. Camden hatte vor allem das Ziel, den Anteil der Büroinvestitionen zurückzudrängen und einen hohen Anteil von Wohnnutzungen zu realisieren. Dieses wurde von den Investoren, British Rail und von Stanhope akzeptiert. Darüber hinaus forderte Camden, daß ein größerer Teil der Wohnungen der Kommune als Sozialwohnung übereignet werden sollten. Hier kam es niemals zu einer völligen Einigung, obwohl die Investoren und Stanhope die Forderung im Grundsatz akzeptieren. Im Vergleich zur Entwicklung von Broadgate zeigt sich hier, daß die Kommune sehr viel selbstbewußter gegenüber den Investoren auftrat. Zwar wurden die gesamten Planungsleistungen und die konzeptionellen Entwicklungen von den privaten Gesellschaften geleistet. Die Kommune übernahm mehr die Rolle eines Korrektors bzw. die Rolle eines Maklers für die Interessen der Bürger in der Umgebung. Diese atypische Rollenverteilung wurde von den Investoren als akzeptabel angesehen. Das Projekt ist nicht an den Forderungen der Kommune sondern seit 1988/89 verschlechterten Investitionsbedingungen für Immobilieninvestitionen gescheitert. Nach Auskunft von den zuständigen Boroughs ist beabsichtigt, sobald sich die ungünstige Marktsituation verbessert hat, mit den Investitionen zu beginnen. Zu diesem Zeitpunkt müssen dann auch die offenen Planungsfragen mit den Kommunen gelöst werden.

Beziehungen zu British Rail

British Rail ist bei den Anträgen auf «planning consent» als Mitantragsteller aufgetreten. D.h., als Partner bei der künftigen Entwicklung. Gleichzeitig bemüht sich British Rail darum, die erheblichen Eisenbahninvestitionen zu realisieren und zu finanzieren. Die Endstation für die Tunneleisenbahn muß nicht aus den örtlichen Grundstückserlösen finanziert werden sondern wird Teil des gesamten Investitionsprojekts betrachtet.

Vergleich von Canary Wharf zu Broadgate

Das Konzept ist noch stärker auf die Bedürfnisse der umgebenden Boroughs bezogen als Broadgate. Neben einem Bürokomplex und einem Platz, der stark auf den Bahnhof bezogen ist entsteht ein Mischgebiet und anschließend reine Wohngebiete. Die Bürozonen weisen eine hohe Dichte (GFZ 5,0) auf. Die Wohngebiete passen sich in Dichte, Höhe und Bauform an die umgebende typische Londonbebauung an. Auch der Stadtgrundriß entspricht eher historischen Entwicklungsformen. Verglichen mit den früheren Projekten kann man eine Abstufung der Beteiligung der Kommunen erkennen. Während Canary Wharf de facto eine autonome Entwicklung von Investoren und einer staatlichen Gesellschaft war, bildet Broadgate eine weitgehend reine Büroentwicklung, die allerdings sehr geschickt in ein bestehendes Gebiet «eingepaßt» wurde. Der Borough hatte im wesentlichen den Vorteil der steigenden Zahl der Arbeitsplätze und der erhöhten Steuereinnahmen bzw. erhöhten Einkaufsmöglichkeiten. In Kings Cross entsteht eine Entwicklung, die noch stärker in die Umgebung integriert ist. Obwohl auch hier der Anstoß zur Investition von British Rail und einer privaten Entwicklungsgesellschaft kam.

Exkurs: Die Rolle des British Rail Property Board
Der British Rail Property Board ist eine Tochterorganisation von British Rail. Sie verwaltet die Grundvermögen von British Rail, vermarktet «Air Rights» und berät den Vorstand von British Rail bei allen Immobilienentwicklungen. British Rail Property Board (BRPB) darf nicht wie eine reine Grundstücksaktiengesellschaft agieren. British Rail darf im Prinzip Grundstücke nur für die Verkehrszwecke erwerben. British Rail kann also keine allgemeine Grundstücksbevorratung betreiben und allgemeine Grundstücksentwicklungen vornehmen. British Rail hat in den 80er Jahren in verschiedenen Bereichen erfolgreiche Entwicklungen von bisherigen Eisenbahnland angestoßen. Das berühmteste Beispiel ist Broadgate. British Rail hat diese Grundstücke zum Teil selbst entwickelt.

Im Prinzip müssen alle öffentlichen Stellen in Großbritannien ein «Register of Public Land» halten. Aus dem ist die Art der Nutzung bezogen auf den jeweiligen Lagewert der Grundstücke zu ersehen. Im Prinzip kann die Regierung öffentliche Körperschaften, die über untergenutztes Land verfügen, zwingen, dieses Land zu verkaufen. Gegenüber British Rail wurde diese Politik nie durchgesetzt, da British Rail immer die Möglichkeit hat, auf künftige Eisenbahnplanungen zu verweisen oder andere wichtige mit der Eisenbahn verknüpfte Investitionsvorhaben aus der Schublade zu ziehen. Die Entwicklung von Broadgate war in weiten Bereichen eine Eisenbahnüberbauung. In Kings Cross wird ebenfalls eine Kombination aus Büro- und Wohnbau mit Weiterentwicklung des Transportsystems geplant. Grundsätzlich werden jedoch Grundstücke, die nicht für Eisenbahnzwecke gebraucht werden, veräußert. British Rail bleibt also seinem engeren Zweck treu. Am Beispiel von Kings Cross zeigt sich jedoch, daß diese Position sehr weit interpretiert werden kann.

Die Entwicklung von Stockley Park

Vorgeschichte

Das Gelände von Stockley Park war eine Hausmüllkippe (160 ha) in der Nähe der M4-Ausfahrt zum Flughafen Heathrow. Stanhope entdeckte «die Chance», dieses flughafen- und autobahnnahe Grundstück, das erhebliche negative Ausstrahlungen auf die angrenzenden Gebiete hatte (Ratten, Gestank, Brände), mitten im Greenbelt zu entwickeln.

Private Finanzierung und Planning
Gain-Investitionen für die Kommunen

Für einen privaten Developer stellte das Projekt eine erhebliche Herausforderung dar, da die Bewältigung der Müllprobleme mit hohen Startkosten verbunden waren. Der gesamte Müll wurde auf einen Hügel (Golf) zusammengetragen und abgedeckt. Wegen der weiterlaufenden Verrottungsprozesse und der Gas- und Abwasserentwicklung waren komplizierte technische Lösungen notwendig (Verwendung der ausströmenden Gase zur Herstellung von Elektrizität, Ableitung und Reinigung der Abwässer, damit Reinigung eines bisher völlig konterminierten Baches). Erst nach einer Lösung dieser Aufgaben konnte mit der Bebauung begonnen werden.

Dabei forderten die Kommunen eine Beteiligung am Entwicklungsgewinn, den Stanhope und die übrigen beteiligten Investoren (u.a. Prudential Versicherung) zu erzielen hofften. Die Developer verpflichteten sich, der Kommune einen Golfplatz kostenlos zur Verfügung zu stellen.

Bebauungskonzepte

Grundsätzlich standen mehrere Bebauungslösungen zur Wahl. Zusammen mit den politischen Körperschaften entschied sich Stanhope für eine lockere und niedriggeschossige Bebauung (drei Stockwerke, GFZ etwa 0,3 in den bebauten Gebieten). Gleichzeitig konzentrierte sich das Angebot auch auf Industrieverwaltungen und Forschungs- und Managementbereiche verschiedener großer Unternehmen. Auf der Grundlage einer Marktanalyse wurde entschieden, ein Arbeiten in einer Parkatmosphäre anzubieten. Die Kunden wollten im Kontrast zu den Arbeitsplätzen in der Kernstadt und insbesondere in der Londoner City locker bebaute, durchgrünte Flächen. Der Anteil der Parkplätze ist hoch. Auf 25 m² Nutzfläche wird ein Parkplatz angeboten. Keines der Gebäude sollte höher sein als der künftige Baumbestand. Damit sollte sich der Park optimal in die Landschaft einfügen.

Investoren und Management

Stanhope blieb in Teilbereichen Eigentümer des Geländes und der Gebäude und tritt damit als der größte Vermieter in Stockley Park auf. Stanhope bleibt gleichzeitig die Managementgesellschaft für den gesamten Park. Zum Teil haben andere Investoren Gebäude errichtet und vermieten diese. Nutzer haben allerdings auch Gebäude für sich selbst entwickelt. Als Koordinator der Entwicklung, der daneben auch für die Konzeption verantwortlich ist, hat Stanhope auf ein sehr einheitliches Erscheinungsbild hingewirkt. Ganz im Gegensatz zu der London Docklands Development Corporation hat Stanhope eigene Gestaltungsvorstellungen, die zusammen mit bekannten Architekten entwickelt wurden, durchgesetzt. Dadurch entsteht heute das Bild einer großen Geschlossenheit, Ruhe und Stimmigkeit. Insgesamt wurde ein High-Tech-Image angestrebt. Die Gebäude wirken leicht und sind weitgehend in hellen Farben gehalten. Um eine möglichst kostengünstige Bebauung und wirtschaftliche Nutzung zu erreichen, wurden im Vergleich zu deutschen Bautraditionen sehr tiefe Gebäude errichtet. Dadurch werden trotz geringer Geschoßzahlen relativ hohe effektive Ausnutzungsziffern erreicht.

Bewertung

Die private Entwicklung des Gebietes wa nur aufgrund seiner extrem günstigen Lage ohne Subventionen möglich. Allerdings zeigt sich, daß es durchaus sinnvoll sein kann, günstige Lagen von privaten Investoren entwickeln zu lassen, während die öffentliche Hand sich auf die schwierigen Lagen, in denen noch Subventionen erforderlich sind, konzentriert.

- Die Planungs- und Gestaltungsvorgaben durch die Entwicklungsgesellschaft haben sich auf die Attraktivität des Parkes sehr günstig ausgewirkt. In einer Periode, in der die Materialien, Bauformen und Ausdrucksformen in der Architektur explodieren, erweist es sich als sehr sinnvoll, durch Vorgaben Eingrenzungen zu schaffen.
- Durch ein einheitliches Management wird der Park in gleicher Qualität bewirtschaftet. Die Außenanlagen werden aus einer Hand gestaltet und gepflegt. Auch dies trägt zum attraktiven Erscheinungsbild des gesamten Parkes bei.
- Die Entwicklung war absolut nachfrageorientiert. Die geringe Baudichte entspricht den Wünschen der Nutzer. Hier stellt sich die Frage, ob in so attraktiver Lage nicht höhere Ausnutzungsziffern sinnvoll gewesen wären, ohne daß sich dabei die Erträge für die Investoren vermindert hätten.
- Es ist zu erwarten, daß sich Entwicklungen dieser Art, z.B. im Umland von Berlin, wiederholen werden. In Westdeutschland dürften private Entwickler allerdings kaum so attraktive Grundstücke finden, die sie in ähnlicher Form bebauen können.

NOTES

1. Vgl. hierzu auch The IPD Property Investment Digest 1990, Investment Property Databank, London 1990
2. Die offenen Immobilienfonds verwalten lediglich ein Vermögen von rund 16 Mrd.DM. Die Versicherungen halten ein hohes Immobilienvermögen in der Größenordnung von 45 Mrd.DM, das allerdings zu Buchwerten bewertet wird und deshalb in seinen Marktwerten kaum abgeschätzt werden kann.
3. Der Stand der Pensionszusagen durch Unternehmen in der Bundesrepublik erreicht rd. 400 Mrd.DM. In den Bilanzen der Großunternehmen machen Pensionszusagen nach Angaben der Bundesbank etwa 15% aus. Etwa 80% der Pensionszusagen treten am Kapitalmarkt nicht in Erscheinung sonder verbleiben den Unternehmen zur Selbstfinanzierung. Dementsprechend halten in der Bundesrepublik die Pensionsfonds nur 12% aller Aktien, in Großbritannien sind es rd. 30%. Entsprechend niedrig dürften in der Bundesrepublik auch die Anlagen in Immobilienfonds sein, deren Anlagevolumen insgesamt nur 16 Mrd.DM erreicht. Damit führen die günstigen steuerlichen Regelungen für Unternehmen dazu, daß die Pensionsverpflichtungen anders als in Großbritannien oder in den USA kaum zum Aufbau einer professionellen Immobilienwirtschaft beitragen. Steuerliche Regelungen haben deshalb ein erhebliches Gewicht für die Entwicklung einer Immobilienwirtschaft. Man muß die Regelungen für die Pensionsrückstellungen im Zusammenhang sehen mit den günstigen steuerrechtlichen Regelungen für private Investoren. Das Entstehen einer professionellen Immobilienwirtschaft wird dadurch von zwei Seiten her erschwert.
4. Die unterschiedlichen Traditionen werden bis in die Bewertungsverfahren von Immobilienfonds hinein sichtbar. Während in Großbritannien eine Gesamtrendite (Perfomance) ermittelt wird, die Veränderung des Verkehrswertes und die Nettoerträge miteinander kombiniert, ermitteln die offenen Immobilienfonds ihre Werte in der Bundesrepublik sehr viel vorsichtiger. Die ermittelten Renditen beruhen nicht auf der Verkehrswertentwicklung der Grundstücke sondern auf Ertragswerten, die bezogen auf die Restwertzeit jeweils mit einem konstanten «Immobilienzins» abdiskontiert werden. Auf diese Weise kommt es zu einer Glättung der Wertentwicklung. De facto bestimmen die Mietsteigerungen die Wertsteigerungen. Zinsschwankungen und Schwankungen der Verkehrswerte haben kaum Einfluß auf die so ermittelten Renditen. Auch hier wird deutlich, die Bewertungspraxis orientiert sich an dem Ziel, möglichst stabile langfristige Anlagen zu fördern. Eine hohe Fungibilität der Anlagen ist weniger erwünscht.
5. Vgl. z.B. Richard Barras: «The Office Development Cycle» in Land Development Studies 1989, S. 35–50
6. Diese im Prinzip als Schweinezyklus bekannte Erklärung kann jedoch nicht ausreichen. Im Laufe der Zeit müßten bei rationalem Verhalten Lerneffekte auftreten. Nähere Analysen zeigen, daß die großen Immobilienkrisen jeweils durch historische Besonderheiten und einmalige Ereignisse bestimmt sind.
7. Vgl. hierzu den Artikel von Rob Harris (Stanhope) in der Stadtbauwelt vom 28.3.91, S. 590, der aus den Diskussionen mit dem Autor zu diesem Gutachten entstanden ist.
8. Vgl. «Development Impact Fees and Other Devices», John Delafons, University of California, Berkeley 1990.

273

9. Vgl. hierzu v.a. Russell Schiller: Retail Decentralisation – The Coming of the Third Wave. In: The Planner Vol 72, 1986, No 7, S. 13–15; Ross Davies/ Elisabeth Howard: Issues in Retail Planning within the United Kingdom. In: Built Environment Vol. 14, 1988, S. 7–21

10. Zum Instrument der Enterprise Zones vgl. z.B. M.G. Lloyd: The Continuing Progress of the Enterprise Zone Experiment. In: Planning Outlook, Vol 29, 1986, No 1, S. 9–12

11. Vgl. hierzu z.B. Graham Parker: Tyneside's rising star. In: Chartered Suveyor Weekly, Vol 14, 1986, S. 437–439; ders.: Out of town, and out of this world? In: Chartered Surveyor Weekly, Vol 10, 1985, S. 442; Derek R. Hall: The Metro Centre and Transport Policy. In: R.D. Knowles: Transport Policy & Urban Development. Transport Geography Study Group, Department of Geography. University of Salford 1989, S 106–189

12. Gateshead Metropolitan Borough Council: Metro Centre Information Note, Gateshead 1990

13. Vgl. hierzu Ron McCarthy: Leisure and Retailing – The Case Study of Meadowhall: In: RSA Journal, Feb. 1990; Sheffield City Council: Retail Assessment of the Meadowhall Development Proposal, Revised Study Nov. 1987, Sheffield 1987

14. Welsh Development International: Joint Ventures. Routes to a Manufacturing Base in the UK. ohne Jahr, S. 19.

15. Durch den Local Goverment-Plan and Land Act von 1980 wurde die Regierung berechtigt Urban Development Corporations ins Leben zu rufen. Nach den 141-146 dieses Gesetzes kann der Minister durch eine spezielle Anordnung den Development Corporations das Eigentum an öffentlichen Grundstücken aufgrund eines Parlaments-beschlusses direkt übertagen.

16. Aus «Zeitschrift für Angewandte Geographie 4/90», Seite 29

17. Aus «The Docklands Experiment», Docklands Consultative Commitee, June 90, S. 85

18. Die geplanten Investitionen am Potsdamer Platz entsprechen in ihren Dimensionen durchaus den Canary Wharf Projekten.

19. In Berlin hat die Stadt nicht den Weg beschritten, Schlüsselgrundstücke durch eine kommunale Gesellschaft entwickeln zu lassen, wie der Verkauf der Potsdamer Platz-Grundstücke an Daimler Benz und Sony demonstriert. Damit wurden Großunter-nehmen, die über keinerlei Erfahrung in der Grundstücksentwicklung verfügen, plötzlich in die Rolle von Stadtbauunternehmen – an der prominentesten Stelle, die in der Bundesrepublik derzeit zu bebauen ist – gedrängt bzw. hineinkatapultiert. Gleichzeitig stehen diese Investoren vor der Forderung, eine hochkomplexe Mischnutz-ungszone sowohl städtebaulich, wie auch aus der Sicht der Nutzer, attraktiv neu zu gestalten. Es liegt auf der Hand, daß diese Investoren ihrerseits wiederum private, erfahrene Grundstücksentwicklungsgesellschaften einschalten müssen (Daimler Benz: ECE; Sony noch offen). Damit entsteht ein sehr kompliziertes Zusammenspiel zwischen dem Investor, seinen Projektentwicklern und der Planung auf der Bezirks- und Landesebene. Eine Lösung von Typus «Docklands Corporation» wäre hier einfacher gewesen. Sie hätte darüber hinaus rechtlich zu einer Joint-Venture-Lösung führen können, die für die Stadt mit hoher Wahrscheinlichkeit auch noch zusätzlich höhere Erträge erbracht hätte.

20. Hinweis: Die Investitionsvorhaben am Potsdamer Platz sind mind. doppelt so groß.

21. Zur Lagesituation: Große Nähe zum Westend und zur City
    – Londons bestentwickelter Knotenpunkt für öffentliche Personennahverkehrsmittel (3 Bahnhöfe von Hauptlinien innerhalb von 5 min. Entfernung.

Beste Fernzugverbindung in alle Richtungen Großbritanniens. Hohe Attraktivität durch die Einfassung des Geländes durch einen Kanal. Bisherige Stadtbrache mit geringen Interessen besetzt.

For Product Safety Concerns and Information please contact our EU
representative GPSR@taylorandfrancis.com
Taylor & Francis Verlag GmbH, Kaufingerstraße 24, 80331 München, Germany